GLENCOE MATHEMATICS

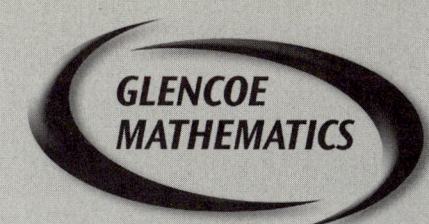

Algebra 1

Lesson Planning Guide

Glencoe McGraw-Hill

New York, New York
Columbus, Ohio
Chicago, Illinois
Peoria, Illinois
Woodland Hills, California

Glencoe/McGraw-Hill

A Division of The McGraw·Hill Companies

Copyright © by The McGraw-Hill Companies, Inc. All rights reserved. Printed in the United States of America. Permission is granted to reproduce the material contained herein on the condition that such material be reproduced only for classroom use; be provided to students, teachers, and families without charge; and be used solely in conjunction with *Glencoe Algebra 1*. Any other reproduction, for use or sale, is prohibited without prior written permission of the publisher.

Send all inquiries to:
Glencoe/McGraw-Hill
8787 Orion Place
Columbus, OH 43240-4027

Algebra 1
Lesson Planning Guide

ISBN: 0-07-827744-2

1 2 3 4 5 6 7 8 9 10 024 07 06 05 04 03 02

CONTENTS

Chapter		Page
	NCTM Principles and Standards	iv
1	The Language of Algebra	1
2	Real Numbers	13
3	Solving Linear Equations	22
4	Graphing Relations and Functions	35
5	Analyzing Linear Equations	47
6	Solving Linear Inequalities	58
7	Solving Systems of Linear Equations and Inequalities	67
8	Polynomials	75
9	Factoring	88
10	Quadratic and Exponential Functions	97
11	Radical Expressions and Triangles	109
12	Rational Expressions and Equations	119
13	Statistics	130
14	Probability	138

NCTM Principles and Standards for School Mathematics

For more information about the National Council of Teachers of Mathematics, log on to www.nctm.org.

NCTM Principles for School Mathematics

- **Equity**   Excellence in mathematics education requires equity—high expectations and strong support for all students.
- **Curriculum**   A curriculum is more than a collection of activities: it must be coherent, focused on important mathematics, and well articulated across the grades.
- **Teaching**   Effective mathematics teaching requires understanding what students know and need to learn and then challenging and supporting them to learn it well.
- **Learning**   Students must learn mathematics with understanding, actively building new knowledge from experience and prior knowledge.
- **Assessment**   Assessment should support the learning of important mathematics and furnish useful information to both teachers and students.
- **Technology**   Technology is essential in teaching and learning mathematics; it influences the mathematics that is taught and enhances students' learning.

NCTM Standards for School Mathematics*

Standard 1:	**Number and Operations**
Standard 2:	**Algebra**
Standard 3:	**Geometry**
Standard 4:	**Measurement**
Standard 5:	**Data Analysis and Probability**
Standard 6:	**Problem Solving**
Standard 7:	**Reasoning and Proof**
Standard 8:	**Communication**
Standard 9:	**Connections**
Standard 10:	**Representation**

* The number assigned to each standard is for easy reference and is not part of each standard's official title.

Lesson Planning Guide (pp. 6–10)

1-1

Teacher's Name _____ Dates _____
Grade _____ Class _____ M Tu W Th F

NCTM Standards
1, 2, 3, 6, 8, 9, 10

Recommended Pacing	
Regular Average	Day 1 of 14
Regular Advanced	Optional
Block Average	Day 1 of 6.5
Block Advanced	Optional

Objectives
___ Write mathematical expressions for verbal expressions.
___ Write verbal expressions for mathematical expressions.
___ State/local objectives: _____

1 Focus
Materials/Resources Needed _____

___ *5-Minute Check Transparencies,* Lesson 1-1
___ *Mathematical Background, TWE,* p. 4C
___ *Prerequisite Skills Masters,* pp. 9–10
___ *TeacherWorks CD-ROM*

2 Teach
___ *In-Class Examples, TWE,* pp. 6–7
___ *Interactive Chalkboard CD-ROM,* Lesson 1-1
___ algebra1.com/extra_examples
___ algebra1.com/career_choices
___ algebra1.com/data_update
___ algebra1.com/usa_today
___ *Guide to Daily Intervention,* pp. 8–9
___ *Daily Intervention, TWE,* p. 7
___ *Study Guide and Intervention, CRM,* pp. 1–2
___ *Reading to Learn Mathematics, CRM,* p. 5
___ *TeacherWorks CD-ROM*
___ *Reading Mathematics, SE,* p. 10

3 Practice/Apply
___ *Skills Practice, CRM,* p. 3
___ *Practice, CRM,* p. 4
___ *Extra Practice, SE,* p. 820
___ *Differentiated Instruction, TWE,* p. 7
___ *Parent and Student Study Guide Workbook,* p. 1
___ *Answer Key Transparencies,* Lesson 1-1
___ algebra1.com/webquest

Assignment Guide, pp. 8–9, SE			
	Objective 1	Objective 2	Other
Basic	11–15 odd, 19, 29, 43, 45	31–37 odd	21–27 odd, 46–57
Average	11–19 odd, 29, 43, 45	31–41 odd	21–27, 46–57
Advanced	12–20 even, 30, 44	32–42 even	22–28 even, 46–49 (optional: 50–57)
Reading Mathematics	1–14		
Alternative Assignment			

4 Assess
___ Open-Ended Assessment, *TWE,* p. 9
___ Enrichment, *CRM,* p. 6
___ algebra1.com/self_check_quiz

___ *Closing the Gap for Absent Students,* pp. 2–3

KEY SE = Student Edition   TWE = Teacher Wraparound Edition   CRM = Chapter Resource Masters

© Glencoe/McGraw-Hill   Glencoe Algebra 1

1-2 Lesson Planning Guide (pp. 11–15)

Teacher's Name _____ Dates _____

Grade _____ Class _____ M Tu W Th F

NCTM Standards
1, 2, 6, 8, 9, 10

Recommended Pacing	
Regular Average	Day 2 of 14
Regular Advanced	Optional
Block Average	Day 1 of 6.5
Block Advanced	Optional

Objectives
- ___ Evaluate numerical expressions by using the order of operations.
- ___ Evaluate algebraic expressions by using the order of operations.
- ___ State/local objectives: _____

1 Focus
Materials/Resources Needed _____

- ___ *5-Minute Check Transparencies*, Lesson 1-2
- ___ *Mathematical Background*, *TWE*, p. 4C
- ___ *Prerequisite Skills Masters*, pp. 5–12, 21–24
- ___ *TeacherWorks CD-ROM*

2 Teach
- ___ In-Class Examples, *TWE*, p. 12
- ___ *Teaching Algebra with Manipulatives*, p. 30
- ___ *Interactive Chalkboard CD-ROM*, Lesson 1-2
- ___ algebra1.com/extra_examples
- ___ *Guide to Daily Intervention*, pp. 8–9
- ___ *Daily Intervention*, *TWE*, p. 13
- ___ *Study Guide and Intervention*, *CRM*, pp. 7–8
- ___ *Reading to Learn Mathematics*, *CRM*, p. 11
- ___ *TeacherWorks CD-ROM*

3 Practice/Apply
- ___ *Skills Practice*, *CRM*, p. 9
- ___ *Practice*, *CRM*, p. 10
- ___ *Extra Practice*, *SE*, p. 820
- ___ *Differentiated Instruction*, *TWE*, p. 15
- ___ *School-to-Career Masters*, p. 1
- ___ *Graphing Calculator and Spreadsheet Masters*, pp. 23, 51
- ___ *Parent and Student Study Guide Workbook*, p. 2
- ___ *Answer Key Transparencies*, Lesson 1-2
- ___ *AlgePASS CD-ROM*, Lesson 1

Assignment Guide, pp. 13–15, SE			
	Objective 1	Objective 2	Other
Basic	15–25 odd, 31	29	30, 33–37 odd, 44–47 51–71
Average	15–27 odd,	29, 33–39 odd	41–47, 51–71 (optional: 48-50)
Advanced	16–28 even, 40	32–48 even 43	30, 41, 42, 44–63 (optional: 64-71)
Alternative Assignment			

4 Assess
- ___ Open-Ended Assessment, *TWE*, p. 15
- ___ Enrichment, *CRM*, p. 12
- ___ algebra1.com/self_check_quiz

___ *Closing the Gap for Absent Students*, pp. 2–3

KEY *SE* = Student Edition   *TWE* = Teacher Wraparound Edition   *CRM* = Chapter Resource Masters

© Glencoe/McGraw-Hill   Glencoe Algebra 1

Lesson Planning Guide (pp. 16–20)

1-3

Teacher's Name _____ Dates _____
Grade _____ Class _____ M Tu W Th F

NCTM Standards
1, 2, 6, 8, 9, 10

Recommended Pacing	
Regular Average	Days 3 & 4 of 14
Regular Advanced	Optional
Block Average	Day 2 of 6.5
Block Advanced	Optional

Objectives
___ Solve open sentence equations.
___ Solve open sentence inequalities.
___ State/local objectives: _____

1 Focus
Materials/Resources Needed _____

___ *5-Minute Check Transparencies,* Lesson 1-3
___ *Mathematical Background, TWE,* p. 4C
___ *Prerequisite Skills Masters,* pp. 1–2, 5–12, 25–26, 48–50, 55–58, 61–62
___ *TeacherWorks CD-ROM*

2 Teach
___ In-Class Examples, *TWE,* p. 17
___ *Interactive Chalkboard CD-ROM,* Lesson 1-3
___ algebra1.com/extra_examples
___ *Guide to Daily Intervention,* pp. 8–9
___ *Study Guide and Intervention, CRM,* pp. 13–14
___ *Reading to Learn Mathematics, CRM,* p. 17
___ *TeacherWorks CD-ROM*

3 Practice/Apply
___ Skills Practice, *CRM,* p. 15
___ Practice, *CRM,* p. 16
___ Extra Practice, *SE,* p. 820
___ Differentiated Instruction, *TWE,* p. 17
___ *Science and Mathematics Lab Manual,* pp. 31–34
___ *Graphing Calculator and Spreadsheet Masters,* p. 24
___ *Parent and Student Study Guide Workbook,* p. 3
___ *Answer Key Transparencies,* Lesson 1-3

Assignment Guide, pp. 18–20, *SE*			
	Objective 1	Objective 2	Other
Basic	15–25 odd, 29–33 odd, 45, 46	26–28	49–67
Average	15–25 odd, 29–35 odd	26–28, 37–43 odd	45, 46, 49–67
Advanced	14–24 even, 30–36 even, 45, 46	38–44 even, 47, 48	49–59 (optional: 60–67)
All	Practice Quiz 1 (1–10)		
Alternate Assignment			

4 Assess
___ Practice Quiz 1, *SE,* p. 20
___ Open-Ended Assessment, *TWE,* p. 20
___ Enrichment, *CRM,* p. 18
___ Assessment, Quiz, *CRM,* p. 69
___ algebra1.com/self_check_quiz

___ *Closing the Gap for Absent Students,* pp. 2–3

KEY *SE* = Student Edition   *TWE* = Teacher Wraparound Edition   *CRM* = Chapter Resource Masters

© Glencoe/McGraw-Hill   Glencoe Algebra 1

1-4 Lesson Planning Guide (pp. 21–25)

Teacher's Name _____ Dates _____

Grade _____ Class _____ M Tu W Th F

NCTM Standards
1, 2, 6, 8, 9, 10

Recommended Pacing	
Regular Average	Day 5 of 14
Regular Advanced	Optional
Block Average	Day 2 of 6.5
Block Advanced	Optional

Objectives
____ Recognize the properties of identity and equality.
____ Use the properties of identity and equality.
____ State/local objectives: _____

1 Focus
Materials/Resources Needed _____

____ *5-Minute Check Transparencies,* Lesson 1-4
____ *Mathematical Background, TWE,* p. 4C
____ *Prerequisite Skills Masters,* pp. 5–8, 11–12, 21–22, 25–26, 55–56
____ *TeacherWorks CD-ROM*

2 Teach
____ In-Class Examples, *TWE,* p. 22
____ *Interactive Chalkboard CD-ROM,* Lesson 1-4
____ algebra1.com/extra_examples
____ algebra1.com/data_update
____ *Guide to Daily Intervention,* pp. 8–9
____ *Study Guide and Intervention, CRM,* pp. 19–20
____ *Reading to Learn Mathematics, CRM,* p. 23
____ *TeacherWorks CD-ROM*

3 Practice/Apply
____ *Skills Practice, CRM,* p. 21
____ *Practice, CRM,* p. 22
____ *Extra Practice, SE,* p. 821
____ *Differentiated Instruction, TWE,* p. 22
____ *Parent and Student Study Guide Workbook,* p. 4
____ *Answer Key Transparencies,* Lesson 1-4

Assignment Guide, pp. 23–25, SE			
	Objective 1	Objective 2	Other
Basic	13–21 odd, 37–40	25–29 odd, 31	30, 32, 33, 44–62
Average	13–23 odd, 37–40	25–29 odd, 31	30, 32, 33, 44–62
Advanced	12–22 even, 37–40	24–28 even, 36	34, 35, 51–56 (optional: 57–62)

Alternate Assignment _____

4 Assess
____ Open-Ended Assessment, *TWE,* p. 25
____ Enrichment, *CRM,* p. 24
____ algebra1.com/self_check_quiz

____ *Closing the Gap for Absent Students,* pp. 2–3

KEY *SE* = Student Edition   *TWE* = Teacher Wraparound Edition   *CRM* = Chapter Resource Masters

© Glencoe/McGraw-Hill   Glencoe Algebra 1

Lesson Planning Guide (pp. 26–31)

1-5

Teacher's Name _____ Dates _____
Grade _____ Class _____ M Tu W Th F

NCTM Standards
1, 2, 6, 8, 9, 10

Recommended Pacing	
Regular Average	Day 6 of 14
Regular Advanced	Optional
Block Average	Day 3 of 6.5
Block Advanced	Optional

Objectives
____ Use the Distributive Property to evaluate expressions.
____ Use the Distributive Property to simplify expressions.
____ State/local objectives: _____

1 Focus
Materials/Resources Needed _____
_____
____ *5-Minute Check Transparencies,* Lesson 1-5
____ *Mathematical Background, TWE,* p. 4D
____ *Prerequisite Skills Masters,* pp. 49–50, 55–56, 77–78
____ *TeacherWorks CD-ROM*

2 Teach
____ In-Class Examples, *TWE,* pp. 27–28
____ *Teaching Algebra with Manipulatives,* pp. 30–32
____ *Interactive Chalkboard CD-ROM,* Lesson 1-5
____ algebra1.com/extra_examples
____ algebra1.com/usa_today
____ *Guide to Daily Intervention,* pp. 8–9
____ *Study Guide and Intervention, CRM,* pp. 25–26
____ *Reading to Learn Mathematics, CRM,* p. 29
____ *TeacherWorks CD-ROM*

3 Practice/Apply
____ Skills Practice, *CRM,* p. 27
____ Practice, *CRM,* p. 28
____ Extra Practice, *SE,* p. 821
____ Differentiated Instruction, *TWE,* p. 29
____ *Parent and Student Study Guide Workbook,* p. 5
____ *Answer Key Transparencies,* Lesson 1-5

Assignment Guide, pp. 29–31, *SE*			
	Objective 1	Objective 2	Other
Basic	15–17 odd, 29, 30, 31–35 odd	19–25 odd, 43–51 odd	54–72
Average	15–17 odd, 29, 30, 31–35 odd, 37, 38	15–27 odd, 43–53 odd	54–72
Advanced	16–18 even, 32–36 even, 37–41	20–28 even, 43–53 even	54–69 (optional: 70–72)
Alternate Assignment			

4 Assess
____ Open-Ended Assessment, *TWE,* p. 31
____ Enrichment, *CRM,* p. 30
____ Assessment, Mid-Chapter Test, *CRM,* p. 71
____ Assessment, Quiz, *CRM,* p. 69
____ algebra1.com/self_check_quiz

____ *Closing the Gap for Absent Students,* pp. 2–3

KEY	*SE* = Student Edition	*TWE* = Teacher Wraparound Edition	*CRM* = Chapter Resource Masters

© Glencoe/McGraw-Hill — Glencoe Algebra 1

Lesson Planning Guide (pp. 32–36)

1-6

Teacher's Name _____ Dates _____

Grade _____ Class _____ M Tu W Th F

NCTM Standards
1, 2, 6, 8, 9, 10

Recommended Pacing	
Regular Average	Day 7 of 14
Regular Advanced	Optional
Block Average	Day 3 of 6.5
Block Advanced	Optional

Objectives
____ Recognize the Commutative and Associative Properties.
____ Use the Commutative and Associative Properties to simplify expressions.
____ State/local objectives: _____

1 Focus
Materials/Resources Needed _____

____ *5-Minute Check Transparencies*, Lesson 1-6
____ Mathematical Background, *TWE*, p. 4D
____ *Prerequisite Skills Masters*, pp. 49–50, 77–78
____ *TeacherWorks CD-ROM*

2 Teach
____ In-Class Examples, *TWE*, pp. 33–34
____ *Interactive Chalkboard CD-ROM*, Lesson 1-6
____ algebra1.com/extra_examples
____ *Guide to Daily Intervention*, pp. 8–9
____ *Study Guide and Intervention*, *CRM*, pp. 31-32
____ *Reading to Learn Mathematics*, *CRM*, p. 35
____ *TeacherWorks CD-ROM*

3 Practice/Apply
____ *Skills Practice*, *CRM*, p. 33
____ *Practice*, *CRM*, p. 34
____ *Extra Practice*, *SE*, p. 821
____ *Differentiated Instruction*, *TWE*, p. 33
____ *Parent and Student Study Guide Workbook*, p. 6
____ *Answer Key Transparencies*, Lesson 1-6
____ *AlgePASS CD-ROM*, Lesson 2

Assignment Guide, pp. 34-36, *SE*			
	Objective 1	Objective 2	Other
Basic	45, 47	17–27 odd, 28, 29, 33–41 odd	48–66
Average	45, 47	17–27 odd, 28, 29, 33–43 odd	48–66
Advanced	46	18–26 even, 30, 31, 32–42 even	48–61 (optional: 62–66)
All	Practice Quiz 2 (1–10)		
Alternate Assignment	_____		

4 Assess
____ Practice Quiz 2, *SE*, p. 36
____ Open-Ended Assessment, *TWE*, p. 36
____ *Enrichment*, *CRM*, p. 36
____ algebra1.com/self_check_quiz

____ *Closing the Gap for Absent Students*, pp. 2–3

KEY *SE* = Student Edition   *TWE* = Teacher Wraparound Edition   *CRM* = Chapter Resource Masters

© Glencoe/McGraw-Hill   Glencoe Algebra 1

Lesson Planning Guide (pp. 37–42)

1-7

Teacher's Name _____ Dates _____

Grade _____ Class _____ M Tu W Th F

NCTM Standards
6, 7, 8, 9, 10

Recommended Pacing	
Regular Average	Day 8 of 14
Regular Advanced	Optional
Block Average	Day 4 of 6.5
Block Advanced	Optional

Objectives
___ Identify the hypothesis and conclusion in a conditional statement.
___ Use a counterexample to show that an assertion is false.
___ State/local objectives: _____

1 Focus
Materials/Resources Needed _____

___ *5-Minute Check Transparencies*, Lesson 1-7
___ *Mathematical Background*, *TWE*, p. 4D
___ *TeacherWorks CD-ROM*

2 Teach
___ *In-Class Examples*, *TWE*, pp. 38–39
___ *Teaching Algebra with Manipulatives*, p. 34
___ *Interactive Chalkboard CD-ROM*, Lesson 1-7
___ algebra1.com/extra_examples
___ *Guide to Daily Intervention*, pp. 8–9
___ *Study Guide and Intervention*, *CRM*, pp. 37–38
___ *Reading to Learn Mathematics*, *CRM*, p. 41
___ *TeacherWorks CD-ROM*

3 Practice/Apply
___ Skills Practice, *CRM*, p. 39
___ Practice, *CRM*, p. 40
___ Extra Practice, *SE*, p. 822
___ Differentiated Instruction, *TWE*, p. 38
___ *Parent and Student Study Guide Workbook*, p. 7
___ *Real-World Transparency and Master*
___ *Answer Key Transparencies*, Lesson 1-7

Assignment Guide, pp. 39–42, *SE*			
	Objective 1	Objective 2	Other
Basic	19–35 odd, 44	45, 50	37–43 odd, 51–78
Average	19–35 odd, 44	37–43 odd, 45, 50	51–78
Advanced	18–34 even, 47, 48	36–42 even, 49, 50	46, 51–72 (optional: 73–78)
Alternate Assignment	_____		

4 Assess
___ Open-Ended Assessment, *TWE*, p. 42
___ Enrichment, *CRM*, p. 42
___ Assessment, Quiz, *CRM*, p. 70
___ algebra1.com/self_check_quiz

___ *Closing the Gap for Absent Students*, pp. 2–3

KEY   *SE* = Student Edition   *TWE* = Teacher Wraparound Edition   *CRM* = Chapter Resource Masters

© Glencoe/McGraw-Hill   Glencoe Algebra 1

Lesson Planning Guide (pp. 43–48)

1-8

Teacher's Name _____ Dates _____

Grade _____ Class _____ M Tu W Th F

NCTM Standards
1, 2, 6, 8, 9, 10

Recommended Pacing	
Regular Average	Day 9 of 14
Regular Advanced	Optional
Block Average	Day 4 of 6.5
Block Advanced	Optional

Objectives
___ Interpret graphs of functions.
___ Draw graphs of functions.
___ State/local objectives: _____

---

1 Focus
Materials/Resources Needed _____
___ Building on Prior Knowledge, *TWE*, p. 43
___ *5-Minute Check Transparencies*, Lesson 1-8
___ Mathematical Background, *TWE*, p. 4D
___ *Prerequisite Skills Masters*, pp. 95–96
___ *TeacherWorks CD-ROM*

2 Teach
___ In-Class Examples, *TWE*, pp. 44–45
___ *Interactive Chalkboard CD-ROM*, Lesson 1-8
___ algebra1.com/extra_examples
___ *Guide to Daily Intervention*, pp. 8–9
___ Study Guide and Intervention, *CRM*, pp. 43–44
___ Reading to Learn Mathematics, *CRM*, p. 47
___ *TeacherWorks CD-ROM*
___ *Multimedia Applications Masters*

3 Practice/Apply
___ Skills Practice, *CRM*, p. 45
___ Practice, *CRM*, p. 46
___ Extra Practice, *SE*, p. 822
___ Differentiated Instruction, *TWE*, p. 45
___ *Parent and Student Study Guide Workbook*, p. 8
___ *Real-World Transparency and Master*
___ *Answer Key Transparencies*, Lesson 1-8
___ *WebQuest and Projects Resources*, p. 21

Assignment Guide, pp. 46–48, *SE*

	Objective 1	Objective 2	Other
Basic	11, 13, 24	15, 21, 22	14, 16, 23, 25–32
Average	11, 13, 24	15, 21, 22	14, 16, 23, 25–32
Advanced	24	18, 22	10, 12, 17, 19, 20, 23, 25–31 (optional: 32)

Alternate Assignment _____

4 Assess
___ Open-Ended Assessment, *TWE*, p. 48
___ Enrichment, *CRM*, p. 48
___ algebra1.com/self_check_quiz

___ *Closing the Gap for Absent Students*, pp. 2–3

KEY *SE* = Student Edition   *TWE* = Teacher Wraparound Edition   *CRM* = Chapter Resource Masters

Algebra Activity (p. 49)
A Follow-Up of Lesson 1-8

1

Teacher's Name _____ Dates _____

Grade _____ Class _____ M Tu W Th F

NCTM Standards
1, 6, 8, 9, 10

Recommended Pacing	
Regular Average	Day 10 of 14
Regular Advanced	Optional
Block Average	Day 4 of 6.5
Block Advanced	Optional

Objectives
____ Use grid paper to investigate real-world functions.
____ State/local objectives: _____

Getting Started
Materials/Resources Needed __grid paper__ _____

Teach
____ *Teaching Algebra with Manipulatives*, p. 35
____ *Glencoe Mathematics Classroom Manipulative Kit*
____ Teaching Strategy, *TWE*, p. 49

Assignment Guide, p. 49, *SE*	
All	1–5
Alternate Assignment	

Assess
____ Study Notebook, *TWE*, p. 49

KEY	*SE* = Student Edition	*TWE* = Teacher Wraparound Edition	*CRM* = Chapter Resource Masters

© Glencoe/McGraw-Hill — Glencoe Algebra 1

1-9 Lesson Planning Guide (pp. 50–55)

Teacher's Name _____ Dates _____

Grade _____ Class _____ M Tu W Th F

NCTM Standards
1, 5, 6, 8, 9, 10

Recommended Pacing	
Regular Average	Day 11 of 14
Regular Advanced	Optional
Block Average	Day 5 of 6.5
Block Advanced	Optional

Objectives
___ Analyze data given in tables and graphs (bar, line, and circle).
___ Determine whether graphs are misleading.
___ State/local objectives: _____

1 Focus
Materials/Resources Needed _____
___ *5-Minute Check Transparencies,* Lesson 1-9
___ *Mathematical Background, TWE,* p. 4D
___ *TeacherWorks CD-ROM*

2 Teach
___ In-Class Examples, *TWE,* pp. 51–52
___ *Interactive Chalkboard CD-ROM,* Lesson 1-9
___ algebra1.com/extra_examples
___ algebra1.com/careers
___ *Guide to Daily Intervention,* pp. 8–9
___ Daily Intervention, *TWE,* p. 51
___ Study Guide and Intervention, *CRM,* pp. 49–50
___ Reading to Learn Mathematics, *CRM,* p. 53
___ *TeacherWorks CD-ROM*

3 Practice/Apply
___ Skills Practice, *CRM,* p. 51
___ Practice, *CRM,* p. 52
___ Extra Practice, *SE,* p. 822
___ Differentiated Instruction, *TWE,* p. 52
___ School-to-Career Masters, p. 2
___ Parent and Student Study Guide Workbook, p. 9
___ *Answer Key Transparencies,* Lesson 1-9
___ algebra1.com/webquest

Assignment Guide, pp. 53–55, *SE*			
	Objective 1	Objective 2	Other
Basic	12, 13, 18	17	19–28
Average	12–15, 18	17	19–28
Advanced	14, 15, 18	16	19–28
Alternate Assignment			

4 Assess
___ Open-Ended Assessment, *TWE,* p. 55
___ Enrichment, *CRM,* p. 54
___ Assessment, Quiz, *CRM,* p. 55
___ algebra1.com/self_check_quiz

___ *Closing the Gap for Absent Students,* pp. 2–3

KEY *SE* = Student Edition   *TWE* = Teacher Wraparound Edition   *CRM* = Chapter Resource Masters

Spreadsheet Investigation (p. 56)
A Follow-Up of Lesson 1-9

Teacher's Name _____ Dates _____

Grade _____ Class _____ M Tu W Th F

NCTM Standards
1, 5, 8, 9, 10

Recommended Pacing	
Regular Average	Day 12 of 14
Regular Advanced	Optional
Block Average	Day 5 of 6.5
Block Advanced	Optional

Objectives
____ Use a computer spreadsheet to display data in different ways.
____ State/local objectives: _____

Getting Started
Materials/Resources Needed computer, spreadsheet software
_____
_____

Teach

Assignment Guide, p. 56, *SE*	
All	1–3
Alternate Assignment	

Assess

KEY *SE* = Student Edition   *TWE* = Teacher Wraparound Edition   *CRM* = Chapter Resource Masters

© Glencoe/McGraw-Hill — Glencoe Algebra 1

Review and Testing (pp. 57–65)

1

Teacher's Name _____ Dates _____

Grade _____ Class _____ M Tu W Th F

Recommended Pacing	
Regular Average	Days 13 & 14 of 14
Regular Advanced	Days 1, 2, 3, & 4 of 4
Block Average	Days 6 & 6.5 of 6.5
Block Advanced	Day 1 of 1

Assess

____ *Parent and Student Study Guide Workbook*, p. 10
____ Vocabulary and Concept Check, *SE*, p. 57
____ Vocabulary Test, *CRM*, p. 68
____ Lesson-by-Lesson Review, *SE*, pp. 57–62
____ Practice Test, *SE*, p. 63
____ Chapter 1 Tests, *CRM*, pp. 55–66
____ Open-Ended Assessment, *CRM*, p. 67
____ Standardized Test Practice, *SE*, pp. 64–65
____ Standardized Test Practice, *CRM*, pp. 73–74
____ Cumulative Review, *CRM*, p. 72
____ *Vocabulary PuzzleMaker CD-ROM*
____ algebra1.com/vocabulary_review
____ algebra1.com/chapter_test
____ algebra1.com/standardized_test
____ *MindJogger Videoquizzes VHS*

Other Assessment Materials

- *TestCheck and Worksheet Builder CD-ROM*

KEY   *SE* = Student Edition   *TWE* = Teacher Wraparound Edition   *CRM* = Chapter Resource Masters

Lesson Planning Guide (pp. 68–72)

2-1

Teacher's Name _____ Dates _____
Grade _____ Class _____ M Tu W Th F

NCTM Standards
1, 6, 8, 9, 10

Recommended Pacing	
Regular Average	Day 1 of 11
Regular Advanced	Optional
Block Average	Day 1 of 6
Block Advanced	Optional

Objectives
____ Graph rational numbers on a number line.
____ Find absolute values of rational numbers.
____ State/local objectives: _____

1 Focus
Materials/Resources Needed _____
____ Building on Prior Knowledge, *TWE*, p. 68
____ *5-Minute Check Transparencies*, Lesson 2-1
____ Mathematical Background, *TWE*, p. 66C
____ *Prerequisite Skills Masters*, pp. 1–4, 15–16, 19–20, 45–46, 55–56, 63–66, 75–76
____ *TeacherWorks CD-ROM*

2 Teach
____ In-Class Examples, *TWE*, pp. 69–70
____ *Interactive Chalkboard CD-ROM*, Lesson 2-1
____ algebra1.com/extra_examples
____ algebra1.com/careers
____ *Guide to Daily Intervention*, pp. 10–11
____ Daily Intervention, *TWE*, p. 70
____ Study Guide and Intervention, *CRM*, pp. 75–76
____ Reading to Learn Mathematics, *CRM*, p. 79
____ *TeacherWorks CD-ROM*

3 Practice/Apply
____ Skills Practice, *CRM*, p. 77
____ Practice, *CRM*, p. 78
____ Extra Practice, *SE*, p. 823
____ Differentiated Instruction, *TWE*, p. 72
____ *Science and Mathematics Lab Manual*, pp. 35–38
____ *Parent and Student Study Guide Workbook*, p. 11
____ Answer Key Transparencies, Lesson 2-1

Assignment Guide, pp. 70–72, *SE*			
	Objective 1	Objective 2	Other
Basic	19–33 odd, 42	35–41 odd, 43–57 odd	44, 60–77
Average	19–33 odd, 42	35–41 odd, 43–57 odd	44, 60–77
Advanced	18–32 even, 58	34–40 even, 46–56 even, 57, 59	60–69 (optional: 70–77)
Alternate Assignment			

4 Assess
____ Open-Ended Assessment, *TWE*, p. 72
____ Enrichment, *CRM*, p. 80
____ algebra1.com/self_check_quiz

____ *Closing the Gap for Absent Students*, pp. 4–5

KEY   *SE* = Student Edition   *TWE* = Teacher Wraparound Edition   *CRM* = Chapter Resource Masters

© Glencoe/McGraw-Hill   Glencoe Algebra 1

Lesson Planning Guide (pp. 73–78)

2-2

Teacher's Name _____ Dates _____

Grade _____ Class _____ M Tu W Th F

NCTM Standards
1, 6, 8, 9, 10

Recommended Pacing	
Regular Average	Days 2 & 3 of 11
Regular Advanced	Optional
Block Average	Days 1 & 2 of 6
Block Advanced	Optional

Objectives
___ Add integers and rational numbers.
___ Subtract integers and rational numbers.
___ State/local objectives: _____

1 Focus
Materials/Resources Needed _____
_____
___ *Building on Prior Knowledge*, TWE, p. 73
___ *5-Minute Check Transparencies*, Lesson 2-2
___ *Mathematical Background*, TWE, p. 73
___ *Prerequisite Skills Masters*, pp. 15–16, 19–24, 39–40, 55–60, 65–66, 75–76
___ *TeacherWorks CD-ROM*

2 Teach
___ *In-Class Examples*, TWE, pp. 74–75
___ *Teaching Algebra with Manipulatives*, pp. 39–41
___ *Interactive Chalkboard CD-ROM*, Lesson 2-2
___ algebra1.com/extra_examples
___ algebra1.com/careers
___ algebra1.com/data_updates
___ *Guide to Daily Intervention*, pp. 10–11
___ *Study Guide and Intervention*, CRM, pp. 81–82
___ *Reading to Learn Mathematics*, CRM, p. 85
___ *TeacherWorks CD-ROM*
___ *Multimedia Applications Masters*

3 Practice/Apply
___ *Skills Practice*, CRM, p. 83
___ *Practice*, CRM, p. 84
___ *Extra Practice*, SE, p. 823
___ *Differentiated Instruction*, TWE, p. 75
___ *Graphing Calculator and Spreadsheet Masters*, p. 26
___ *Parent and Student Study Guide Workbook*, p. 12
___ *Answer Key Transparencies*, Lesson 2-2
___ *WebQuest and Projects Resources*, p. 21

Assignment Guide, pp. 76–78, SE			
	Objective 1	Objective 2	Other
Basic	17–33 odd, 37	39–53 odd	57–59, 63–82
Average	17–37 odd	39–55 odd	57–59, 63–82
Advanced	18–38 even	40–56 even	60–76 (optional: 77–82)

Alternate Assignment _____

4 Assess
___ *Open-Ended Assessment*, TWE, p. 78
___ *Enrichment*, CRM, p. 86
___ *Assessment, Quiz*, CRM, p. 131
___ algebra1.com/self_check_quiz

___ *Closing the Gap for Absent Students*, pp. 4–5

KEY SE = Student Edition TWE = Teacher Wraparound Edition CRM = Chapter Resource Masters

© Glencoe/McGraw-Hill *Glencoe Algebra 1*

Lesson Planning Guide (pp. 79–83)

2-3

Teacher's Name _____ Dates _____
Grade _____ Class _____ M Tu W Th F

NCTM Standards
1, 6, 8, 9, 10

Recommended Pacing	
Regular Average	Day 4 of 11
Regular Advanced	Optional
Block Average	Day 2 of 6
Block Advanced	Optional

Objectives
____ Multiply integers.
____ Multiply rational numbers.
____ State/local objectives: _____

1 Focus

Materials/Resources Needed _____
_____
____ *5-Minute Check Transparencies*, Lesson 2-3
____ *Mathematical Background*, *TWE*, p. 66C
____ *Prerequisite Skills Masters*, pp. 15–16, 19–20, 25–28, 39–40, 47–50, 65–66, 75–76
____ *TeacherWorks CD-ROM*

2 Teach
____ In-Class Examples, *TWE*, p. 80
____ *Teaching Algebra with Manipulatives*, p. 42
____ *Interactive Chalkboard CD-ROM*, Lesson 2-3
____ algebra1.com/extra_examples
____ algebra1.com/usa_today
____ *Guide to Daily Intervention*, pp. 10–11
____ *Study Guide and Intervention*, *CRM*, pp. 87–88
____ *Reading to Learn Mathematics*, *CRM*, p. 91
____ *TeacherWorks CD-ROM*

3 Practice/Apply
____ Skills Practice, *CRM*, p. 89
____ Practice, *CRM*, p. 90
____ Extra Practice, *SE*, p. 823
____ Differentiated Instruction, *TWE*, p. 81
____ *School-to-Career Masters*, p. 3
____ *Parent and Student Study Guide Workbook*, p. 13
____ *Real-World Transparency and Master*
____ *Answer Key Transparencies*, Lesson 2-3

Assignment Guide, pp. 81–83, *SE*			
	Objective 1	Objective 2	Other
Basic	17–21 odd, 35–39 odd	23–31 odd, 41	40, 43–47 odd, 51, 55–76
Average	17–21 odd	41–47 odd, 51	23–39 odd, 40, 49, 55–76
Advanced	16–20 even, 34–38 even	22–32 even, 42–50 even	52–68 (optional: 69–76)
All	Practice Quiz 1 (1–10)		
Alternate Assignment	_____		

4 Assess
____ Practice Quiz 1, *SE*, p. 83
____ Open-Ended Assessment, *TWE*, p. 83
____ Enrichment, *CRM*, p. 92
____ algebra1.com/self_check_quiz

____ *Closing the Gap for Absent Students*, pp. 4–5

KEY *SE* = Student Edition   *TWE* = Teacher Wraparound Edition   *CRM* = Chapter Resource Masters

© Glencoe/McGraw-Hill   Glencoe Algebra 1

Lesson Planning Guide (pp. 84–87)

2-4

Teacher's Name _____ Dates _____

Grade _____ Class _____ M Tu W Th F

NCTM Standards
1, 6, 8, 9, 10

Recommended Pacing	
Regular Average	Day 5 of 11
Regular Advanced	Optional
Block Average	Day 3 of 6
Block Advanced	Optional

Objectives

___ Divide integers.
___ Divide rational numbers.
___ State/local objectives: _____

1 Focus

Materials/Resources Needed _____

___ *5-Minute Check Transparencies,* Lesson 2-4
___ *Mathematical Background, TWE,* p. 66D
___ *Prerequisite Skills Masters,* pp. 15–16, 17–20, 29–32, 39–40, 47–48, 51–52, 53–54, 63–66, 75–76
___ *TeacherWorks CD-ROM*

2 Teach

___ *In-Class Examples, TWE,* p. 85
___ *Teaching Algebra with Manipulatives,* pp. 43–44
___ *Interactive Chalkboard CD-ROM,* Lesson 2-4
___ algebra1.com/extra_examples
___ *Guide to Daily Intervention,* pp. 10–11
___ *Study Guide and Intervention, CRM,* pp. 93–94
___ *Reading to Learn Mathematics, CRM,* p. 97
___ *TeacherWorks CD-ROM*

3 Practice/Apply

___ *Skills Practice, CRM,* p. 95
___ *Practice, CRM,* p. 96
___ *Extra Practice, SE,* p. 824
___ *Differentiated Instruction, TWE,* p. 85
___ *School-to-Career Masters,* p. 4
___ *Parent and Student Study Guide Workbook,* p. 14
___ *Answer Key Transparencies,* Lesson 2-4
___ *AlgePASS CD-ROM,* Lesson 3

Assignment Guide, pp. 85–87, SE			
	Objective 1	Objective 2	Other
Basic	17, 37–41 odd	19–31 odd, 35, 45–51 odd	55, 58–77
Average	17, 37–43 odd	19–35 odd	45–55 odd, 58–77
Advanced	18, 38–44 even	20–36 even, 46–54 even	56–73 (optional: 74–77)
Alternate Assignment			_____

4 Assess

___ Open-Ended Assessment, *TWE,* p. 86
___ *Enrichment, CRM,* p. 98
___ Assessment, Mid-Chapter Test, *CRM,* p. 133
___ Assessment, Quiz, *CRM,* p. 131
___ algebra1.com/self_check_quiz

___ *Closing the Gap for Absent Students,* pp. 4–5

KEY *SE* = Student Edition *TWE* = Teacher Wraparound Edition *CRM* = Chapter Resource Masters

© Glencoe/McGraw-Hill   Glencoe Algebra 1

Lesson Planning Guide (pp. 88–95)

2-5

Teacher's Name _____ Dates _____
Grade _____ Class _____ M Tu W Th F

NCTM Standards
1, 5, 6, 8, 9, 10

Recommended Pacing	
Regular Average	Day 6 of 11
Regular Advanced	Optional
Block Average	Days 3 & 4 of 6
Block Advanced	Optional

Objectives
___ Interpret and create line plots and stem-and-leaf plots.
___ Analyze data using mean, median, and mode.
___ State/local objectives: _____

1 Focus

Materials/Resources Needed _____

___ Building on Prior Knowledge, *TWE*, p. 88
___ *5-Minute Check Transparencies,* Lesson 2-5
___ Mathematical Background, *TWE*, p. 88
___ *Prerequisite Skills Masters,* pp. 15–16, 61–62, 75–76
___ *TeacherWorks CD-ROM*

2 Teach

___ In-Class Examples, *TWE*, pp. 89–91
___ *Teaching Algebra with Manipulatives,* pp. 45–47
___ *Interactive Chalkboard CD-ROM,* Lesson 2-5
___ algebra1.com/extra_examples
___ *Guide to Daily Intervention,* pp. 10–11
___ Daily Intervention, *TWE*, p. 91
___ Study Guide and Intervention, *CRM*, pp. 99–100
___ Reading to Learn Mathematics, *CRM*, p. 103
___ *TeacherWorks CD-ROM*
___ Reading Mathematics, *SE*, p. 95

3 Practice/Apply

___ Skills Practice, *CRM*, p. 101
___ Practice, *CRM*, p. 102
___ Extra Practice, *SE*, p. 824
___ Differentiated Instruction, *TWE*, p. 90
___ *Parent and Student Study Guide Workbook,* p. 15
___ *Answer Key Transparencies,* Lesson 2-5

Assignment Guide, pp. 91–94, *SE*			
	Objective 1	Objective 2	Other
Basic	15, 16, 21, 22	17–19, 23–25, 27	26, 38, 42–64
Average	15, 21, 28, 32	25, 27, 29–31, 33, 34, 39–41	26, 38, 42–64
Advanced	14, 20, 35	36, 37, 39–41	38, 42–56 (optional: 57–64)
Reading Mathematics	1–6		
Alternate Assignment			

4 Assess

___ Open-Ended Assessment, *TWE*, p. 94
___ Enrichment, *CRM*, p. 104
___ algebra1.com/self_check_quiz

___ *Closing the Gap for Absent Students,* pp. 4–5

KEY *SE* = Student Edition   *TWE* = Teacher Wraparound Edition   *CRM* = Chapter Resource Masters

Lesson Planning Guide (pp. 96–101)

2-6

Teacher's Name _____ Dates _____

Grade _____ Class _____ M Tu W Th F

NCTM Standards
1, 5, 6, 7, 8, 9, 10

Recommended Pacing	
Regular Average	Day 7 of 11
Regular Advanced	Optional
Block Average	Day 4 of 6
Block Advanced	Optional

Objectives
____ Find the probability of a simple event.
____ Find the odds of a simple event.
____ State/local objectives: _____

1 Focus
Materials/Resources Needed _____
_____
____ *5-Minute Check Transparencies*, Lesson 2-6
____ *Mathematical Background*, *TWE*, p. 66D
____ *Prerequisite Skills Masters*, pp. 17–18, 37–38, 67–70, 99–100
____ *TeacherWorks CD-ROM*

2 Teach
____ *In-Class Examples*, *TWE*, pp. 97–98
____ *Teaching Algebra with Manipulatives*, p. 48
____ *Interactive Chalkboard CD-ROM*, Lesson 2-6
____ algebra1.com/extra_examples
____ *Guide to Daily Intervention*, pp. 10–11
____ *Daily Intervention*, *TWE*, p. 97
____ *Study Guide and Intervention*, *CRM*, pp. 105–106
____ *Reading to Learn Mathematics*, *CRM*, p. 109
____ *TeacherWorks CD-ROM*

3 Practice/Apply
____ *Skills Practice*, *CRM*, p. 107
____ *Practice*, *CRM*, p. 108
____ *Extra Practice*, *SE*, p. 824
____ *Differentiated Instruction*, *TWE*, p. 98
____ *Parent and Student Study Guide Workbook*, p. 16
____ *Answer Key Transparencies*, Lesson 2-6
____ algebra1.com/webquest

Assignment Guide, pp. 99–100, SE			
	Objective 1	Objective 2	Other
Basic	15–35 odd, 51	37–49 odd, 53	52, 59–82
Average	15–31 odd, 51	37–49 odd, 52, 53, 55, 56	54, 59–82
Advanced	14–34 even, 54, 57	36–50 even, 55, 56, 58	59–74 (optional: 75–82)
All	Practice Quiz 2 (1–10)		
Alternate Assignment			

4 Assess
____ Practice Quiz 2, *SE*, p. 101
____ Open-Ended Assessment, *TWE*, p. 101
____ Enrichment, *CRM*, p. 110
____ Assessment, Quiz, *CRM*, p. 132
____ algebra1.com/self_check_quiz

____ *Closing the Gap for Absent Students*, pp. 4–5

KEY *SE* = Student Edition   *TWE* = Teacher Wraparound Edition   *CRM* = Chapter Resource Masters

© Glencoe/McGraw-Hill   Glencoe Algebra 1

Algebra Activity (p. 102)
A Follow-Up of Lesson 2-6

Teacher's Name _____ Dates _____

Grade _____ Class _____ M Tu W Th F

NCTM Standards
1, 5, 6, 7, 8, 9, 10

Recommended Pacing	
Regular Average	Day 8 of 11
Regular Advanced	Optional
Block Average	Day 5 of 6
Block Advanced	Optional

Objectives
___ Use tables to investigate probability and Pascal's Triangle.
___ State/local objectives: _____

Getting Started
Materials/Resources Needed _____
_____
_____

Teach
___ *Teaching Algebra with Manipulatives*, p. 49
___ *Glencoe Mathematics Classroom Manipulative Kit*
___ Teaching Strategy, *TWE*, p. 102

Assignment Guide, p. 102, SE	
All	1–7
Alternate Assignment	

Assess
___ Study Notebook, *TWE*, p. 102

KEY *SE* = Student Edition   *TWE* = Teacher Wraparound Edition   *CRM* = Chapter Resource Masters

© Glencoe/McGraw-Hill   Glencoe Algebra 1

Lesson Planning Guide (pp. 103–109)

2-7

Teacher's Name _____ Dates _____

Grade _____ Class _____ M Tu W Th F

NCTM Standards
1, 6, 8, 9, 10

Recommended Pacing	
Regular Average	Day 9 of 11
Regular Advanced	Optional
Block Average	Day 5 of 6
Block Advanced	Optional

Objectives
___ Find square roots.
___ Classify and order real numbers.
___ State/local objectives: _____

---

1 Focus

Materials/Resources Needed _____

___ Building on Prior Knowledge, *TWE*, p. 103
___ 5-Minute Check Transparencies, Lesson 2-7
___ Mathematical Background, *TWE*, p. 66D
___ Prerequisite Skills Masters, pp. 75–76
___ TeacherWorks CD-ROM

2 Teach

___ In-Class Examples, *TWE*, pp. 104–106
___ Interactive Chalkboard CD-ROM, Lesson 2-7
___ algebra1.com/extra_examples
___ Guide to Daily Intervention, pp. 10–11
___ Study Guide and Intervention, *CRM*, pp. 111–112
___ Reading to Learn Mathematics, *CRM*, p. 115
___ TeacherWorks CD-ROM

3 Practice/Apply

___ Skills Practice, *CRM*, p. 113
___ Practice, *CRM*, p. 114
___ Extra Practice, *SE*, p. 825
___ Differentiated Instruction, *TWE*, p. 107
___ Graphing Calculator and Spreadsheet Masters, p. 25
___ Parent and Student Study Guide Workbook, p. 17
___ Answer Key Transparencies, Lesson 2-7

Assignment Guide, pp. 107–109, *SE*

	Objective 1	Objective 2	Other
Basic	21–31 odd	33–47 odd, 53, 55, 59, 61, 65–69 odd	51, 73, 77–88
Average	21–31 odd	33–49 odd, 53–69 odd	51, 70–73, 77–88
Advanced	20–30 even, 50	32–48 even, 52–68 even	73–88

Alternate Assignment _____

4 Assess

___ Open-Ended Assessment, *TWE*, p. 109
___ Enrichment, *CRM*, p. 116
___ Assessment, Quiz, *CRM*, p. 132
___ algebra1.com/self_check_quiz

___ Closing the Gap for Absent Students, pp. 4–5

KEY *SE* = Student Edition *TWE* = Teacher Wraparound Edition *CRM* = Chapter Resource Masters

© Glencoe/McGraw-Hill   Glencoe Algebra 1

Review and Testing (pp. 110–117)

2

Teacher's Name _____ Dates _____

Grade _____ Class _____ M Tu W Th F

Recommended Pacing	
Regular Average	Days 10 & 11 of 11
Regular Advanced	Days 1, 2, 3, & 4 of 4
Block Average	Day 6 of 6
Block Advanced	Day 1 of 1

Assess
____ *Parent and Student Study Guide Workbook*, p. 18
____ Vocabulary and Concept Check, *SE*, p. 110
____ Vocabulary Test, *CRM*, p. 130
____ Lesson-by-Lesson Review, *SE*, pp. 110–114
____ Practice Test, *SE*, p. 115
____ Chapter 2 Tests, *CRM*, pp. 117–128
____ Open-Ended Assessment, *CRM*, p. 129
____ Standardized Test Practice, *SE*, pp. 116–117
____ Standardized Test Practice, *CRM*, pp. 135–136
____ Cumulative Review, *CRM*, p. 134
____ *Vocabulary PuzzleMaker CD-ROM*
____ algebra1.com/vocabulary_review
____ algebra1.com/chapter_test
____ algebra1.com/standardized_test
____ *MindJogger Videoquizzes VHS*

Other Assessment Materials
- *TestCheck and Worksheet Builder CD-ROM*

KEY   SE = Student Edition   TWE = Teacher Wraparound Edition   CRM = Chapter Resource Masters

© Glencoe/McGraw-Hill   Glencoe Algebra 1

Lesson Planning Guide (pp. 120–126)

3-1

Teacher's Name _____ Dates _____

Grade _____ Class _____ M Tu W Th F

NCTM Standards
1, 2, 6, 8, 9, 10

Recommended Pacing	
Regular Average	Day 1 of 16
Regular Advanced	Optional
Block Average	Day 1 of 8
Block Advanced	Optional

Objectives
____ Translate verbal sentences into equations.
____ Translate equations into verbal sentences.
____ State/local objectives: _____

1 Focus
Materials/Resources Needed _____

____ *5-Minute Check Transparencies*, Lesson 3-1
____ *Mathematical Background, TWE*, p. 118C
____ *TeacherWorks CD-ROM*

2 Teach
____ In-Class Examples, *TWE*, pp. 121–123
____ *Teaching Algebra with Manipulatives*, pp. 56–58
____ *Interactive Chalkboard CD-ROM*, Lesson 3-1
____ algebra1.com/extra_examples
____ *Guide to Daily Intervention*, pp. 12–13
____ *Study Guide and Intervention, CRM*, pp. 137–138
____ *Reading to Learn Mathematics, CRM*, p. 141
____ *TeacherWorks CD-ROM*

3 Practice/Apply
____ Skills Practice, *CRM*, p. 139
____ Practice, *CRM*, p. 140
____ Extra Practice, *SE*, p. 825
____ Differentiated Instruction, *TWE*, p. 121
____ *Parent and Student Study Guide Workbook*, p. 19
____ *Answer Key Transparencies*, Lesson 3-1

Assignment Guide, pp. 123–126, *SE*			
	Objective 1	Objective 2	Other
Basic	13–17 odd, 21–27 odd, 41–47, 52	29–39 odd	53–76
Average	13–27 odd, 41–47, 52	29–39 odd	53–76
Advanced	14–28 even, 43–51	30–40 even	53–68 (optional: 69–76)
Alternate Assignment			

4 Assess
____ Open-Ended Assessment, *TWE*, p. 126   ____ algebra1.com/self_check_quiz
____ Enrichment, *CRM*, p. 142

____ *Closing the Gap for Absent Students*, pp. 6–7

KEY   *SE* = Student Edition   *TWE* = Teacher Wraparound Edition   *CRM* = Chapter Resource Masters

© Glencoe/McGraw-Hill   Glencoe Algebra 1

Algebra Activity (p. 127)
A Preview of Lesson 3-2

3

Teacher's Name _____ Dates _____

Grade _____ Class _____ M Tu W Th F

NCTM Standards
1, 2, 10

Recommended Pacing	
Regular Average	Day 2 of 16
Regular Advanced	Optional
Block Average	Day 1 of 8
Block Advanced	Optional

Objectives
___ Use algebra tiles to solve addition and subtraction equations.
___ State/local objectives: _____

Getting Started
Materials/Resources Needed equation mats, algebra tiles

Teach
___ *Teaching Algebra with Manipulatives,* p. 59
___ *Glencoe Mathematics Classroom Manipulative Kit*
___ Teaching Strategy, *TWE,* p. 127

Assignment Guide, p. 127, *SE*	
All	1–8
Alternate Assignment	

Assess
___ Study Notebook, *TWE,* p. 127

KEY *SE* = Student Edition   *TWE* = Teacher Wraparound Edition   *CRM* = Chapter Resource Masters

Lesson Planning Guide (pp. 128–134)

3-2

Teacher's Name _____ Dates _____

Grade _____ Class _____ M Tu W Th F

NCTM Standards
1, 2, 6, 8, 9, 10

Recommended Pacing	
Regular Average	Day 3 of 16
Regular Advanced	Optional
Block Average	Day 2 of 8
Block Advanced	Optional

Objectives
___ Solve equations by using addition.
___ Solve equations by using subtraction.
___ State/local objectives: _____

---

1 Focus
Materials/Resources Needed _____

___ Building on Prior Knowledge, *TWE*, p. 128
___ *5-Minute Check Transparencies*, Lesson 3-2
___ Mathematical Background, *TWE*, p. 118C
___ *Prerequisite Skills Masters*, pp. 21–22, 59–60
___ *TeacherWorks CD-ROM*

2 Teach
___ In-Class Examples, *TWE*, pp. 129–131
___ *Teaching Algebra with Manipulatives*, pp. 60–65
___ *Interactive Chalkboard CD-ROM*, Lesson 3-2
___ algebra1.com/extra_examples
___ *Guide to Daily Intervention*, pp. 12–13
___ Daily Intervention, *TWE*, p. 129
___ Study Guide and Intervention, *CRM*, pp. 143–144
___ Reading to Learn Mathematics, *CRM*, p. 147
___ *TeacherWorks CD-ROM*

3 Practice/Apply
___ Skills Practice, *CRM*, p. 145
___ Practice, *CRM*, p. 146
___ Extra Practice, *SE*, p. 825
___ Differentiated Instruction, *TWE*, p. 130
___ *Parent and Student Study Guide Workbook*, p. 20
___ *Answer Key Transparencies*, Lesson 3-2

Assignment Guide, pp. 131–134, *SE*			
	Objective 1	Objective 2	Other
Basic	15–37 odd, 41–49 odd, 51–57	15–37 odd, 41–49 odd, 51–57	65–89
Average	15–49 odd, 51–61	15–49 odd, 51–61	65–89
Advanced	16–50 even, 58–64	16–50 even, 58–64	65–81 (optional: 82–89)
Alternate Assignment	_____		

4 Assess
___ Open-Ended Assessment, *TWE*, p. 134
___ Enrichment, *CRM*, p. 148
___ algebra1.com/self_check_quiz

___ Closing the Gap for Absent Students, pp. 6–7

KEY *SE* = Student Edition *TWE* = Teacher Wraparound Edition *CRM* = Chapter Resource Masters

Lesson Planning Guide (pp. 135–140)

3-3

Teacher's Name _____ Dates _____

Grade _____ Class _____ M Tu W Th F

NCTM Standards
1, 2, 6, 8, 9, 10

Recommended Pacing	
Regular Average	Day 4 of 16
Regular Advanced	Optional
Block Average	Days 2 & 3 of 8
Block Advanced	Optional

Objectives
____ Solve equations by using multiplication.
____ Solve equations by using division.
____ State/local objectives: _____

---

1 Focus
Materials/Resources Needed _____

____ Building on Prior Knowledge, *TWE*, p. 136
____ 5-Minute Check Transparencies, Lesson 3-3
____ Mathematical Background, *TWE*, p. 118C
____ Prerequisite Skills Masters, pp. 9–12, 51–52
____ TeacherWorks CD-ROM

2 Teach
____ In-Class Examples, *TWE*, pp. 136–137
____ Teaching Algebra with Manipulatives, pp. 66–67
____ Interactive Chalkboard CD-ROM, Lesson 3-3
____ algebra1.com/extra_examples
____ Guide to Daily Intervention, pp. 12–13
____ Study Guide and Intervention, *CRM*, pp. 149–150
____ Reading to Learn Mathematics, *CRM*, p. 153
____ TeacherWorks CD-ROM

3 Practice/Apply
____ Skills Practice, *CRM*, p. 151
____ Practice, *CRM*, p. 152
____ Extra Practice, *SE*, p. 826
____ Differentiated Instruction, *TWE*, p. 137
____ Parent and Student Study Guide Workbook, p. 21
____ Answer Key Transparencies, Lesson 3-3
____ WebQuest and Projects Resources, p. 22

Assignment Guide, pp. 137–140, *SE*			
	Objective 1	Objective 2	Other
Basic	13–29 odd, 33, 35, 39–41, 43–45	13–29 odd, 33, 35, 39–41, 43–45	50–70
Average	13–37 odd, 39–41, 43–45	13–37 odd, 39–41, 43–45	50–70
Advanced	14–38 even, 43–49	14–38 even, 43–49	50–66 (optional: 67–70)
All	Practice Quiz 1 (1–10)		
Alternate Assignment			

4 Assess
____ Practice Quiz 1, *SE*, p. 140
____ Open-Ended Assessment, *TWE*, p. 140
____ Enrichment, *CRM*, p. 154

____ Assessment, Quiz, *CRM*, p. 205
____ algebra1.com/self_check_quiz

____ Closing the Gap for Absent Students, pp. 6–7

KEY *SE* = Student Edition   *TWE* = Teacher Wraparound Edition   *CRM* = Chapter Resource Masters

© Glencoe/McGraw-Hill   Glencoe Algebra 1

Algebra Activity (p. 141)
A Preview of Lesson 3-4

3

Teacher's Name _____ Dates _____

Grade _____ Class _____ M Tu W Th F

NCTM Standards
1, 2, 10

Recommended Pacing	
Regular Average	Day 5 of 16
Regular Advanced	Optional
Block Average	Day 3 of 8
Block Advanced	Optional

Objectives
___ Use algebra tiles to solve multi-step equations.
___ State/local objectives: _____

Getting Started
Materials/Resources Needed __equation mats, algebra tiles__ _____

Teach
___ *Teaching Algebra with Manipulatives*, p. 68
___ *Glencoe Mathematics Classroom Manipulative Kit*
___ *Teaching Strategy*, *TWE*, p. 141

Assignment Guide, p. 141, *SE*	
All	1–9
Alternate Assignment	

Assess
___ Study Notebook, *TWE*, p. 141

KEY   *SE* = Student Edition   *TWE* = Teacher Wraparound Edition   *CRM* = Chapter Resource Masters

© Glencoe/McGraw-Hill   Glencoe Algebra 1

Lesson Planning Guide (pp. 142–148)

3-4

Teacher's Name _____ Dates _____

Grade _____ Class _____ M Tu W Th F

NCTM Standards
1, 2, 6, 8, 9, 10

Recommended Pacing	
Regular Average	Days 6 & 7 of 16
Regular Advanced	Optional
Block Average	Day 3 of 8
Block Advanced	Optional

Objectives
___ Solve problems by working backward.
___ Solve equations involving more than one operation.
___ State/local objectives: _____

1 Focus
Materials/Resources Needed _____
___ *5-Minute Check Transparencies*, Lesson 3-4
___ *Mathematical Background*, *TWE*, p. 118C
___ *Prerequisite Skills Masters*, pp. 77–78
___ *TeacherWorks CD-ROM*

2 Teach
___ *In-Class Examples*, *TWE*, pp. 143–144
___ *Teaching Algebra with Manipulatives*, pp. 69–72
___ *Interactive Chalkboard CD-ROM*, Lesson 3-4
___ algebra1.com/extra_examples
___ *Guide to Daily Intervention*, pp. 12–13
___ *Study Guide and Intervention*, *CRM*, pp. 155–156
___ *Reading to Learn Mathematics*, *CRM*, p. 159
___ *TeacherWorks CD-ROM*

3 Practice/Apply
___ *Skills Practice*, *CRM*, p. 157
___ *Practice*, *CRM*, p. 158
___ *Extra Practice*, *SE*, p. 826
___ *Differentiated Instruction*, *TWE*, p. 144
___ *School-to-Career Masters*, p. 5
___ *Science and Mathematics Lab Manual*, pp. 39–42
___ *Parent and Student Study Guide Workbook*, p. 22
___ *Answer Key Transparencies*, Lesson 3-4
___ *AlgePASS CD-ROM*, Lesson 4

Assignment Guide, pp. 145—148, *SE*			
	Objective 1	Objective 2	Other
Basic	17–21 odd	23–39 odd, 43–49 odd, 51–53	55–58, 65–89
Average	17–21 odd	23–49 odd, 51–53	55–58, 65–89 (optional: 59–64)
Advanced	16–20 even	22–50 even, 54	55–83 (optional: 84–89)
Alternate Assignment			

4 Assess
___ Open-Ended Assessment, *TWE*, p. 148
___ Enrichment, *CRM*, p. 160
___ algebra1.com/self_check_quiz

___ *Closing the Gap for Absent Students*, pp. 6–7

KEY SE = Student Edition   TWE = Teacher Wraparound Edition   CRM = Chapter Resource Masters

© Glencoe/McGraw-Hill   Glencoe Algebra 1

Lesson Planning Guide (pp. 149–154)

3-5

Teacher's Name _____ Dates _____
Grade _____ Class _____ M Tu W Th F

NCTM Standards
1, 2, 6, 8, 9, 10

Recommended Pacing	
Regular Average	Day 8 of 16
Regular Advanced	Optional
Block Average	Day 4 of 8
Block Advanced	Optional

Objectives
___ Solve equations with the variable on each side.
___ Solve equations involving grouping symbols.
___ State/local objectives: _____

1 Focus
Materials/Resources Needed _____
_____
___ *5-Minute Check Transparencies*, Lesson 3-5
___ *Mathematical Background, TWE*, p. 118D
___ *Prerequisite Skills Masters*, pp. 23–24
___ *TeacherWorks CD-ROM*

2 Teach
___ In-Class Examples, *TWE*, pp. 150–151
___ *Teaching Algebra with Manipulatives*, pp. 73–74
___ *Interactive Chalkboard CD-ROM*, Lesson 3-5
___ algebra1.com/extra_examples
___ *Guide to Daily Intervention*, pp. 12–13
___ Study Guide and Intervention, *CRM*, pp. 161–162
___ Reading to Learn Mathematics, *CRM*, p. 165
___ *TeacherWorks CD-ROM*

3 Practice/Apply
___ Skills Practice, *CRM*, p. 163
___ Practice, *CRM*, p. 164
___ Extra Practice, *SE*, p. 826
___ Differentiated Instruction, *TWE*, p. 150
___ *Graphing Calculator and Spreadsheet Masters*, p. 27
___ *Parent and Student Study Guide Workbook*, p. 23
___ *Answer Key Transparencies*, Lesson 3-5
___ *AlgePASS CD-ROM*, Lesson 5

Assignment Guide, pp. 152–154, *SE*			
	Objective 1	Objective 2	Other
Basic	15–21 odd, 31, 35, 37, 41	23–29 odd, 33, 39, 43, 45	47, 48
Average	15–21 odd, 31, 35, 37, 41	23–29 odd, 33, 39, 43, 45, 47	49–75
Advanced	14–20 even, 30, 34, 36, 40, 46	22–28 even, 32, 38, 42, 44, 48	49–67 (optional: 68–75)
Alternate Assignment			

4 Assess
___ Open-Ended Assessment, *TWE*, p. 154
___ Enrichment, *CRM*, p. 166
___ Assessment, Mid-Chapter Test, *CRM*, p. 207
___ Assessment, Quiz, *CRM*, p. 205
___ algebra1.com/self_check_quiz

___ *Closing the Gap for Absent Students*, pp. 6–7

KEY *SE* = Student Edition   *TWE* = Teacher Wraparound Edition   *CRM* = Chapter Resource Masters

© Glencoe/McGraw-Hill   Glencoe Algebra 1

Lesson Planning Guide (pp. 155–159)

3-6

Teacher's Name _____ Dates _____

Grade _____ Class _____ M Tu W Th F

NCTM Standards
1, 2, 3, 6, 8, 9, 10

Recommended Pacing	
Regular Average	Day 9 of 16
Regular Advanced	Optional
Block Average	Day 4 of 8
Block Advanced	Optional

Objectives
___ Determine whether two ratios form a proportion.
___ Solve proportions.
___ State/local objectives: _____

---

1 Focus
Materials/Resources Needed _____

___ *5-Minute Check Transparencies,* Lesson 3-6
___ *Mathematical Background, TWE,* p. 118D
___ *Prerequisite Skills Masters,* pp. 27–28, 67–74
___ *TeacherWorks CD-ROM*

2 Teach
___ In-Class Examples, *TWE,* pp. 156–157
___ *Teaching Algebra with Manipulatives,* pp. 75–76
___ *Interactive Chalkboard CD-ROM,* Lesson 3-6
___ algebra1.com/extra_examples
___ *Guide to Daily Intervention,* pp. 12–13
___ Study Guide and Intervention, *CRM,* pp. 167–168
___ Reading to Learn Mathematics, *CRM,* p. 171
___ *TeacherWorks CD-ROM*
___ *Multimedia Applications Masters*

3 Practice/Apply
___ Skills Practice, *CRM,* p. 169
___ Practice, *CRM,* p. 170
___ Extra Practice, *SE,* p. 827
___ Differentiated Instruction, *TWE,* p. 157
___ *School-to-Career Masters,* p. 6
___ *Parent and Student Study Guide Workbook,* p. 24
___ *Answer Key Transparencies,* Lesson 3-6
___ *AlgePASS CD-ROM,* Lesson 6
___ *WebQuest and Projects Resources,* p. 29
___ algebra1.com/webquest

Assignment Guide, pp. 157–159, SE			
	Objective 1	Objective 2	Other
Basic	11–17 odd	19–23 odd, 31, 33	36–58
Average	11–17 odd	19–35 odd	36–58
Advanced	12–18 even	20–34 even	36–54 (optional: 55–58)

Alternate Assignment _____

4 Assess
___ Open-Ended Assessment, *TWE,* p. 159
___ Enrichment, *CRM,* p. 172
___ algebra1.com/self_check_quiz

___ *Closing the Gap for Absent Students,* pp. 6–7

KEY SE = Student Edition   TWE = Teacher Wraparound Edition   CRM = Chapter Resource Masters

© Glencoe/McGraw-Hill   Glencoe Algebra 1

Lesson Planning Guide (pp. 160–165)

3-7

Teacher's Name _____ Dates _____

Grade _____ Class _____ M Tu W Th F

NCTM Standards
1, 2, 6, 8, 9, 10

Recommended Pacing	
Regular Average	Day 10 of 16
Regular Advanced	Optional
Block Average	Day 5 of 8
Block Advanced	Optional

Objectives
___ Find percents of increase and decrease.
___ Solve problems involving percents of change.
___ State/local objectives: _____

1 Focus
Materials/Resources Needed _____

___ *5-Minute Check Transparencies,* Lesson 3-7
___ *Mathematical Background, TWE,* p. 118D
___ *Prerequisite Skills Masters,* pp. 17–18, 41–44, 71–72, 77–78
___ *TeacherWorks CD-ROM*

2 Teach
___ In-Class Examples, *TWE,* p. 161
___ *Teaching Algebra with Manipulatives,* pp. 77–79
___ *Interactive Chalkboard CD-ROM,* Lesson 3-7
___ algebra1.com/extra_examples
___ algebra1.com/careers
___ *Guide to Daily Intervention,* pp. 12–13
___ Daily Intervention, *TWE,* p. 162
___ Study Guide and Intervention, *CRM,* pp. 173–174
___ Reading to Learn Mathematics, *CRM,* p. 177
___ *TeacherWorks CD-ROM*
___ Reading Mathematics, *SE,* p. 165

3 Practice/Apply
___ Skills Practice, *CRM,* p. 175
___ Practice, *CRM,* p. 176
___ Extra Practice, *SE,* p. 827
___ Differentiated Instruction, *TWE,* p. 161
___ *Graphing Calculator and Spreadsheet Masters,* p. 28
___ *Parent and Student Study Guide Workbook,* p. 25
___ *Real-World Transparency and Master*
___ *Answer Key Transparencies,* Lesson 3-7
___ *WebQuest and Projects Resources,* p. 29

Assignment Guide, pp. 162–164, SE			
	Objective 1	Objective 2	Other
Basic	15–27 odd	31–41 odd	49–71
Average	15–29 odd, 46, 47	31–45 odd	49–71
Advanced	14–30 even, 46, 48	32–44 even	43–47 odd, 49–65 (optional: 66–71)
All	Practice Quiz 2 (1–10)		
Reading Mathematics	1–3		
Alternate Assignment			

4 Assess
___ Practice Quiz 2, *SE,* p. 164
___ Open-Ended Assessment, *TWE,* p. 164
___ Enrichment, *CRM,* p. 178

___ Assessment, Quiz, *CRM,* p. 206
___ algebra1.com/self_check_quiz

___ *Closing the Gap for Absent Students,* pp. 6–7

KEY *SE* = Student Edition   *TWE* = Teacher Wraparound Edition   *CRM* = Chapter Resource Masters

© Glencoe/McGraw-Hill   Glencoe Algebra 1

Lesson Planning Guide (pp. 166–170)

3-8

Teacher's Name _____ Dates _____

Grade _____ Class _____ M Tu W Th F

NCTM Standards
1, 2, 6, 8, 9, 10

Recommended Pacing	
Regular Average	Days 11 & 12 of 16
Regular Advanced	Optional
Block Average	Days 5 & 6 of 8
Block Advanced	Optional

Objectives
____ Solve equations for given variables.
____ Use formulas to solve real-world problems.
____ State/local objectives: _____

1 Focus
Materials/Resources Needed _____

____ *5-Minute Check Transparencies,* Lesson 3-8
____ *Mathematical Background, TWE,* p. 118D
____ *Prerequisite Skills Masters,* pp. 81–82
____ *TeacherWorks CD-ROM*

2 Teach
____ In-Class Examples, *TWE,* p. 167
____ *Teaching Algebra with Manipulatives,* pp. 80–81
____ *Interactive Chalkboard CD-ROM,* Lesson 3-8
____ algebra1.com/extra_examples
____ *Guide to Daily Intervention,* pp. 12–13
____ *Study Guide and Intervention, CRM,* pp. 179, 180
____ *Reading to Learn Mathematics, CRM,* p. 183
____ *TeacherWorks CD-ROM*
____ *Multimedia Applications Masters*

3 Practice/Apply
____ Skills Practice, *CRM,* p. 181
____ Practice, *CRM,* p. 182
____ Extra Practice, *SE,* p. 827
____ Differentiated Instruction, *TWE,* p. 168
____ *Parent and Student Study Guide Workbook,* p. 26
____ Answer Key Transparencies, Lesson 3-8
____ *AlgePASS CD-ROM,* Lessons 7 & 8
____ *WebQuest and Projects Resources,* p. 22

Assignment Guide, pp. 168–170, SE			
	Objective 1	Objective 2	Other
Basic	13–25 odd	34–39	42–66
Average	13–29 odd	34–39	42–66
Advanced	14–32 even	38–41	31, 33, 42–60 (optional: 61–66)

Alternate Assignment _____

4 Assess
____ Open-Ended Assessment, *TWE,* p. 170
____ Enrichment, *CRM,* p. 184
____ algebra1.com/self_check_quiz

____ *Closing the Gap for Absent Students,* pp. 6–7

KEY *SE* = Student Edition   *TWE* = Teacher Wraparound Edition   *CRM* = Chapter Resource Masters

© Glencoe/McGraw-Hill   Glencoe Algebra 1

Lesson Planning Guide (pp. 171–177)

3-9

Teacher's Name _____ Dates _____
Grade _____ Class _____ M Tu W Th F

NCTM Standards
1, 2, 6, 8, 9, 10

Recommended Pacing	
Regular Average	Day 13 of 16
Regular Advanced	Optional
Block Average	Day 6 of 8
Block Advanced	Optional

Objectives
___ Solve mixture problems.
___ Solve uniform motion problems.
___ State/local objectives: _____

1 Focus
Materials/Resources Needed _____
___ *5-Minute Check Transparencies,* Lesson 3-9
___ *Mathematical Background, TWE,* p. 118D
___ *TeacherWorks CD-ROM*

2 Teach
___ *In-Class Examples, TWE,* pp. 172–173
___ *Interactive Chalkboard CD-ROM,* Lesson 3-9
___ algebra1.com/extra_examples
___ algebra1.com/data_update
___ *Guide to Daily Intervention,* pp. 12–13
___ *Study Guide and Intervention, CRM,* pp. 185–186
___ *Reading to Learn Mathematics, CRM,* p. 189
___ *TeacherWorks CD-ROM*

3 Practice/Apply
___ *Skills Practice, CRM,* p. 187
___ *Practice, CRM,* p. 188
___ *Extra Practice, SE,* p. 828
___ *Differentiated Instruction, TWE,* p. 173
___ *Parent and Student Study Guide Workbook,* p. 27
___ *Answer Key Transparencies,* Lesson 3-9
___ algebra1.com/webquest

Assignment Guide, pp. 174–177, SE			
	Objective 1	Objective 2	Other
Basic	11–14, 23–29 odd	19–21, 31	36–50
Average	11–14, 23–29 odd, 33	19–21, 26, 31	35–50
Advanced	15–18, 22, 24, 28	19–21, 26, 30–32, 34	33, 35–50
Alternate Assignment			

4 Assess
___ Open-Ended Assessment, *TWE,* p. 177
___ Enrichment, *CRM,* p. 190
___ Assessment, Quiz, *CRM,* p. 206
___ algebra1.com/self_check_quiz

___ *Closing the Gap for Absent Students,* pp. 6–7

KEY SE = Student Edition   TWE = Teacher Wraparound Edition   CRM = Chapter Resource Masters

© Glencoe/McGraw-Hill   Glencoe Algebra 1

3 Spreadsheet Investigation (p. 178)
A Follow-Up of Lesson 3-9

Teacher's Name _____ Dates _____

Grade _____ Class _____ M Tu W Th F

NCTM Standards
1, 2, 6, 9

Recommended Pacing	
Regular Average	Day 14 of 16
Regular Advanced	Optional
Block Average	Day 7 of 8
Block Advanced	Optional

Objectives
____ Use a spreadsheet to find a weighted average.
____ State/local objectives: _____

Getting Started

Materials/Resources Needed computer, spreadsheet software

Teach

Assignment Guide, p. 178, *SE*	
All	1–4
Alternate Assignment	

Assess

KEY *SE* = Student Edition   *TWE* = Teacher Wraparound Edition   *CRM* = Chapter Resource Masters

© Glencoe/McGraw-Hill   Glencoe Algebra 1

Review and Testing (pp. 179–187)

3

Teacher's Name _____ Dates _____

Grade _____ Class _____ M Tu W Th F

Recommended Pacing	
Regular Average	Days 15 & 16 of 16
Regular Advanced	Days 1, 2, 3, & 4 of 4
Block Average	Days 7 & 8 of 8
Block Advanced	Day 1 of 1

Assess
___ *Parent and Student Study Guide Workbook,* p. 29
___ Vocabulary and Concept Check, *SE,* p. 179
___ Vocabulary Test, *CRM,* p. 204
___ Lesson-by-Lesson Review, *SE,* pp. 179–184
___ Practice Test, *SE,* p. 185
___ Chapter 3 Tests, *CRM,* pp. 191–202
___ Open-Ended Assessment, *CRM,* p. 203
___ Standardized Test Practice, *SE,* pp. 186–187
___ Standardized Test Practice, *CRM,* pp. 209–210
___ Cumulative Review, *CRM,* p. 208
___ *Vocabulary PuzzleMaker CD-ROM*
___ algebra1.com/vocabulary_review
___ algebra1.com/chapter_test
___ algebra1.com/standardized_test
___ *MindJogger Videoquizzes VHS*
___ Unit 1 Test, *CRM,* pp. 211–212

Other Assessment Materials
- *TestCheck and Worksheet Builder CD-ROM*

KEY *SE* = Student Edition   *TWE* = Teacher Wraparound Edition   *CRM* = Chapter Resource Masters

Lesson Planning Guide (pp. 192–196)

4-1

Teacher's Name _____ Dates _____

Grade _____ Class _____ M Tu W Th F

NCTM Standards
2, 3, 6, 8, 9, 10

Recommended Pacing	
Regular Average	Day 1 of 14
Regular Advanced	Day 1 of 14
Block Average	Day 1 of 7
Block Advanced	Day 1 of 7

Objectives
___ Locate points on the coordinate plane.
___ Graph points on a coordinate plane.
___ State/local objectives: _____

1 Focus
Materials/Resources Needed _____

___ Building on Prior Knowledge, *TWE*, p. 192
___ *5-Minute Check Transparencies*, Lesson 4-1
___ Mathematical Background, *TWE*, p. 190C
___ *TeacherWorks CD-ROM*

2 Teach
___ In-Class Examples, *TWE*, pp. 193–194
___ *Interactive Chalkboard CD-ROM*, Lesson 4-1
___ algebra1.com/extra_examples
___ *Guide to Daily Intervention*, pp. 14–15
___ Study Guide and Intervention, *CRM*, pp. 213–214
___ Reading to Learn Mathematics, *CRM*, p. 217
___ *TeacherWorks CD-ROM*

3 Practice/Apply
___ Skills Practice, *CRM*, p. 215
___ Practice, *CRM*, p. 216
___ Extra Practice, *SE*, p. 828
___ Differentiated Instruction, *TWE*, p. 196
___ *Parent and Student Study Guide Workbook*, p. 29
___ *Answer Key Transparencies*, Lesson 4-1

Assignment Guide, pp. 194–196, *SE*			
	Objective 1	Objective 2	Other
Basic	13–21 odd, 37–43 odd	25–29 odd	31–35 odd, 44–47, 51–71
Average	13–23 odd, 37–43 odd	25–29 odd	31–35 odd, 44–47, 51–71 (optional: 48–50)
Advanced	14–24 even, 38–42 even	26–30 even	32–36 even, 44–66 (optional: 66–71)

Alternate Assignment _____

4 Assess
___ Open-Ended Assessment, *TWE*, p. 196
___ Enrichment, *CRM*, p. 218
___ algebra1.com/self_check_quiz

___ *Closing the Gap for Absent Students*, pp. 8–9

KEY	*SE* = Student Edition	*TWE* = Teacher Wraparound Edition	*CRM* = Chapter Resource Masters

© Glencoe/McGraw-Hill — Glencoe Algebra 1

Lesson Planning Guide (pp. 197–203)

4-2

Teacher's Name _____ Dates _____

Grade _____ Class _____ M Tu W Th F

NCTM Standards
2, 3, 6, 8, 9, 10

Recommended Pacing	
Regular Average	Days 2 & 3 of 14
Regular Advanced	Days 2 & 3 of 14
Block Average	Days 1 & 2 of 7
Block Advanced	Days 1 & 2 of 7

Objectives

___ Transform figures by using reflections, translations, dilations, and rotations.

___ Transform figures on a coordinate plane by using reflections, translations, dilations, and rotations.

___ State/local objectives: _____

1 Focus

Materials/Resources Needed _____
_____

___ Building on Prior Knowledge, *TWE*, p. 197
___ *5-Minute Check Transparencies*, Lesson 4-2
___ Mathematical Background, *TWE*, p. 190C
___ *TeacherWorks CD-ROM*

2 Teach

___ In-Class Examples, *TWE*, pp. 198–200
___ *Interactive Chalkboard CD-ROM*, Lesson 4-2
___ algebra1.com/extra_examples
___ *Guide to Daily Intervention*, pp. 14–15
___ Daily Intervention, *TWE*, p. 198
___ Study Guide and Intervention, *CRM*, pp. 219–220
___ Reading to Learn Mathematics, *CRM*, p. 223
___ *TeacherWorks CD-ROM*

3 Practice/Apply

___ Skills Practice, *CRM*, p. 221
___ Practice, *CRM*, p. 222
___ Extra Practice, *SE*, p. 828
___ Differentiated Instruction, *TWE*, p. 199
___ *Graphing Calculator and Spreadsheet Masters*, p. 29
___ *Parent and Student Study Guide Workbook*, p. 30
___ *Answer Key Transparencies*, Lesson 4-2

Assignment Guide, pp. 200–203, *SE*			
	Objective 1	Objective 2	Other
Basic	11–15 odd, 34	17–23 odd, 27–31 odd, 35, 36	28, 39–43, 47–59
Average	11–15 odd, 34	17–33 odd, 35, 36	39–43, 47–59 (optional: 44–46)
Advanced	12–16 even, 37, 38	18–26 even, 30, 32	39–57 (optional: 58, 59)
Alternate Assignment			

4 Assess

___ Open–Ended Assessment, *TWE*, p. 230
___ Enrichment, *CRM*, p. 224
___ Assessment, Quiz, *CRM*, p. 275
___ algebra1.com/self_check_quiz

___ *Closing the Gap for Absent Students*, pp. 8–9

KEY *SE* = Student Edition   *TWE* = Teacher Wraparound Edition   *CRM* = Chapter Resource Masters

Graphing Calculator Investigation (p. 204)
A Preview of Lesson 4-3

Teacher's Name _____ Dates _____

Grade _____ Class _____ M Tu W Th F

NCTM Standards
3, 8

Recommended Pacing	
Regular Average	Day 4 of 14
Regular Advanced	Day 4 of 14
Block Average	Day 2 of 14
Block Advanced	Day 2 of 14

Objectives
____ Use a graphing calculator to graph relations.
____ State/local objectives: _____

Getting Started
Materials/Resources Needed _____

Teach
____ *Graphing Calculator and Spreadsheet Masters*, p. 52
____ algebra1.com/other_calculator_keystrokes

Assignment Guide, p. 204, SE	
All	1–5
Alternate Assignment	

Assess

KEY   *SE* = Student Edition   *TWE* = Teacher Wraparound Edition   *CRM* = Chapter Resource Masters

© Glencoe/McGraw-Hill   Glencoe Algebra 1

Lesson Planning Guide (pp. 205–211)

4-3

Teacher's Name _____ Dates _____

Grade _____ Class _____ M Tu W Th F

NCTM Standards
2, 3, 6, 8, 9, 10

Recommended Pacing	
Regular Average	Day 5 of 14
Regular Advanced	Day 5 of 14
Block Average	Day 3 of 7
Block Advanced	Day 3 of 7

Objectives

____ Represent relations as sets of ordered pairs, tables, mappings, and graphs.
____ Find the inverse of a relation.
____ State/local objectives: _____

1 Focus

Materials/Resources Needed _____
_____

____ 5-Minute Check Transparencies, Lesson 4-3
____ Mathematical Background, *TWE*, p. 190C
____ TeacherWorks CD-ROM

2 Teach

____ In-Class Examples, *TWE*, pp. 206–207
____ *Teaching Algebra with Manipulatives*, pp. 85–86
____ Interactive Chalkboard CD-ROM, Lesson 4-3
____ algebra1.com/extra_examples
____ algebra1.com/data_update
____ *Guide to Daily Intervention*, pp. 14–15
____ Study Guide and Intervention, *CRM*, pp. 225–226
____ Reading to Learn Mathematics, *CRM*, p. 229
____ TeacherWorks CD-ROM

3 Practice/Apply

____ Skills Practice, *CRM*, p. 227
____ Practice, *CRM*, p. 228
____ Extra Practice, *SE*, p. 829
____ Differentiated Instruction, *TWE*, p. 206
____ School-to-Career Masters, p. 7
____ Science and Mathematics Lab Manual, pp. 47–52
____ Parent and Student Study Guide Workbook, p. 31
____ Answer Key Transparencies, Lesson 4-3
____ WebQuest and Projects Resources, p. 30

Assignment Guide, pp. 208–211, *SE*			
	Objective 1	Objective 2	Other
Basic	19–37 odd, 38	27–37 odd, 39	40, 49–52, 57–80
Average	19–37 odd, 38	27–37 odd, 39	40–43, 49–52, 57–80 (optional: 53–56)
Advanced	18–36 even	26–36 even, 48	41–47, 49–74 (optional: 75–80)
All	Practice Quiz 1 (1–10)		
Alternate Assignment			

4 Assess

____ Practice Quiz 1, *SE*, p. 211
____ Open-Ended Assessment, *TWE*, p. 211
____ Enrichment, *CRM*, p. 230
____ algebra1.com/self_check_quiz

____ Closing the Gap for Absent Students, pp. 8–9

KEY   *SE* = Student Edition   *TWE* = Teacher Wraparound Edition   *CRM* = Chapter Resource Masters

© Glencoe/McGraw-Hill   Glencoe Algebra 1

Lesson Planning Guide (pp. 212–217)

4-4

Teacher's Name _____ Dates _____
Grade _____ Class _____ M Tu W Th F

NCTM Standards
2, 3, 6, 8, 9, 10

Recommended Pacing	
Regular Average	Day 6 of 14
Regular Advanced	Day 6 of 14
Block Average	Day 3 of 7
Block Advanced	Day 3 of 7

Objectives

___ Use an equation to determine the range for a given domain.
___ Graph the solution set for a given domain.
___ State/local objectives: _____

1 Focus

Materials/Resources Needed _____

___ Building on Prior Knowledge, *TWE*, p. 213
___ *5-Minute Check Transparencies*, Lesson 4-4
___ Mathematical Background, *TWE*, p. 190D
___ *TeacherWorks CD-ROM*

2 Teach

___ In-Class Examples, *TWE*, pp. 213–214
___ *Interactive Chalkboard CD-ROM*, Lesson 4-4
___ algebra1.com/extra_examples
___ algebra1.com/career_choices
___ *Guide to Daily Intervention*, pp. 14–15
___ Study Guide and Intervention, *CRM*, pp. 231–232
___ Reading to Learn Mathematics, *CRM*, p. 235
___ *TeacherWorks CD-ROM*

3 Practice/Apply

___ Skills Practice, *CRM*, p. 233
___ Practice, *CRM*, p. 234
___ Extra Practice, *SE*, p. 829
___ Differentiated Instruction, *TWE*, p. 213
___ *Parent and Student Study Guide Workbook*, p. 32
___ *Answer Key Transparencies*, Lesson 4-4

Assignment Guide, pp. 215–217, *SE*			
	Objective 1	Objective 2	Other
Basic	15–37 odd, 40–44 even	33–37 odd	39, 47–51, 56–76
Average	15–37 odd, 40–44 even	33–37 odd, 45	39–43 odd, 47–51, 56–76 (optional: 52–55)
Advanced	14–36 even, 42, 44	32–36 even	38, 43, 46–70 (optional: 71–76)
Alternate Assignment			_____

4 Assess

___ Open-Ended Assessment, *TWE*, p. 217
___ Enrichment, *CRM*, p. 236
___ Assessment, Mid-Chapter Test, *CRM*, p. 277
___ Assessment, Quiz, *CRM*, p. 275
___ algebra1.com/self_check_quiz

___ *Closing the Gap for Absent Students*, pp. 8–9

KEY *SE* = Student Edition   *TWE* = Teacher Wraparound Edition   *CRM* = Chapter Resource Masters

© Glencoe/McGraw-Hill   Glencoe Algebra 1

Lesson Planning Guide (pp. 218–223)

4-5

Teacher's Name _____ Dates _____

Grade _____ Class _____ M Tu W Th F

NCTM Standards
2, 3, 6, 8, 9, 10

Recommended Pacing	
Regular Average	Day 7 of 14
Regular Advanced	Day 7 of 14
Block Average	Day 4 of 7
Block Advanced	Day 4 of 7

Objectives
____ Determine whether an equation is linear.
____ Graph linear equations.
____ State/local objectives: _____

---

1 Focus
Materials/Resources Needed _____
_____

____ *5-Minute Check Transparencies,* Lesson 4-5
____ *Mathematical Background, TWE,* p. 190D
____ *TeacherWorks CD-ROM*

2 Teach
____ *In-Class Examples, TWE,* pp. 219–220
____ *Teaching Algebra with Manipulatives,* p. 87
____ *Interactive Chalkboard CD-ROM,* Lesson 4-5
____ algebra1.com/extra_examples
____ *Guide to Daily Intervention,* pp. 14–15
____ *Daily Intervention, TWE,* p. 219
____ *Study Guide and Intervention, CRM,* pp. 237–238
____ *Reading to Learn Mathematics, CRM,* p. 241
____ *TeacherWorks CD-ROM*

3 Practice/Apply
____ *Skills Practice, CRM,* p. 239
____ *Practice, CRM,* p. 240
____ *Extra Practice, SE,* p. 829
____ *Differentiated Instruction, TWE,* p. 220
____ *Graphing Calculator and Spreadsheet Masters,* p. 30
____ *Parent and Student Study Guide Workbook,* p. 33
____ *Answer Key Transparencies,* Lesson 4-5
____ *AlgePASS CD-ROM,* Lesson 9

Assignment Guide, pp. 221–223, *SE*			
	Objective 1	Objective 2	Other
Basic	17–25 odd	27–41 odd, 48, 50	45–47, 57–84
Average	17–25 odd	27–43 odd, 50	45, 49, 51–53, 57–84
Advanced	16–24 even	26–42 even	44, 52–78 (optional: 79–84)

Alternate Assignment _____

4 Assess
____ Open-Ended Assessment, *TWE,* p. 223
____ Enrichment, *CRM,* p. 242
____ algebra1.com/self_check_quiz

____ *Closing the Gap for Absent Students,* pp. 8–9

KEY	*SE* = Student Edition	*TWE* = Teacher Wraparound Edition	*CRM* = Chapter Resource Masters

Graphing Calculator Investigation (pp. 224-225)

4

Teacher's Name _____ Dates _____

Grade _____ Class _____ M Tu W Th F

NCTM Standards
2, 3, 10

Recommended Pacing	
Regular Average	Day 8 of 14
Regular Advanced	Day 8 of 14
Block Average	Day 4 of 7
Block Advanced	Day 4 of 7

Objectives
___ Use a graphing calculator to graph relations.
___ State/local objectives: _____

Getting Started
Materials/Resources Needed _____
_____
_____
_____

Teach
___ *Graphing Calculator and Spreadsheet Masters*, p. 53
___ algebra1.com/other_calculator_keystrokes

Assignment Guide, p. 225, *SE*	
All	1–15
Alternate Assignment	

Assess

KEY *SE* = Student Edition   *TWE* = Teacher Wraparound Edition   *CRM* = Chapter Resource Masters

© Glencoe/McGraw-Hill   Glencoe Algebra 1

4-6 Lesson Planning Guide (pp. 226–231)

Teacher's Name _____   Dates _____

Grade _____ Class _____ M Tu W Th F

NCTM Standards
2, 3, 6, 8, 9, 10

Recommended Pacing	
Regular Average	Day 9 of 14
Regular Advanced	Day 9 of 14
Block Average	Day 5 of 7
Block Advanced	Day 5 of 7

Objectives
____ Determine whether a relation is a function.
____ Find function values.
____ State/local objectives: _____

1 Focus
Materials/Resources Needed _____
_____
____ *5-Minute Check Transparencies,* Lesson 4-6
____ *Mathematical Background, TWE,* p. 226
____ *TeacherWorks CD-ROM*

2 Teach
____ In-Class Examples, *TWE,* pp. 227–228
____ *Teaching Algebra with Manipulatives,* pp. 88–89
____ *Interactive Chalkboard CD-ROM,* Lesson 4-6
____ algebra1.com/extra_examples
____ *Guide to Daily Intervention,* pp. 14–15
____ Study Guide and Intervention, *CRM,* pp. 243–244
____ Reading to Learn Mathematics, *CRM,* p. 247
____ *TeacherWorks CD-ROM*

3 Practice/Apply
____ Skills Practice, *CRM,* p. 245
____ Practice, *CRM,* p. 246
____ Extra Practice, *SE,* p. 830
____ Differentiated Instruction, *TWE,* p. 227
____ *School-to-Career Masters,* p. 8
____ *Science and Mathematics Lab Manual,* pp. 141–146
____ *Parent and Student Study Guide Workbook,* p. 34
____ *Real-World Transparency and Master*
____ *Answer Key Transparencies,* Lesson 4-6
____ *AlgePASS CD-ROM,* Lesson 10
____ algebra1.com/webquest

Assignment Guide, pp. 229–231, *SE*			
	Objective 1	Objective 2	Other
Basic	17–31 odd	33–41 odd	45–58, 52–70
Average	17–31 odd	33–43 odd	45–58, 52–70
Advanced	18–30 even	32–42 even	44, 49–65 (optional: 65–70)
All	Practice Quiz 2 (1–10)		
Alternate Assignment			

4 Assess
____ Practice Quiz 2, *SE,* p. 231
____ Open-Ended Assessment, *TWE,* p. 231
____ Enrichment, *CRM,* p. 248

____ Assessment, Quiz, *CRM,* p. 276
____ algebra1.com/self_check_quiz

____ *Closing the Gap for Absent Students,* pp. 8–9

KEY	SE = Student Edition	TWE = Teacher Wraparound Edition	CRM = Chapter Resource Masters

© Glencoe/McGraw-Hill   Glencoe Algebra 1

Spreadsheet Investigation (p. 232)
A Preview of Lesson 4-7

4

Teacher's Name _____ Dates _____

Grade _____ Class _____ M Tu W Th F

NCTM Standards
1, 2, 6

Recommended Pacing	
Regular Average	Day 10 of 14
Regular Advanced	Day 10 of 14
Block Average	Day 5 of 7
Block Advanced	Day 5 of 7

Objectives
___ Use a spreadsheet to generate number sequences and patterns.
___ State/local objectives: _____

Getting Started
Materials/Resources Needed _____

Teach

Assignment Guide, p. 232, SE	
All	1–5
Alternate Assignment	

Assess

KEY *SE* = Student Edition   *TWE* = Teacher Wraparound Edition   *CRM* = Chapter Resource Masters

© Glencoe/McGraw-Hill   Glencoe Algebra 1

Lesson Planning Guide (pp. 233–239)

4-7

Teacher's Name _____ Dates _____

Grade _____ Class _____ M Tu W Th F

NCTM Standards
1, 2, 6, 8, 9, 10

Recommended Pacing	
Regular Average	Day 11 of 14
Regular Advanced	Day 11 of 14
Block Average	Day 6 of 7
Block Advanced	Day 6 of 7

Objectives
____ Recognize arithmetic sequences.
____ Extend and write formulas for arithmetic sequences.
____ State/local objectives: _____

1 Focus
Materials/Resources Needed _____
_____
____ Building on Prior Knowledge, *TWE*, p. 234
____ *5-Minute Check Transparencies,* Lesson 4-7
____ Mathematical Background, *TWE*, p. 190D
____ *Prerequisite Skills Masters,* pp. 5–8
____ *TeacherWorks CD-ROM*

2 Teach
____ In-Class Examples, *TWE*, pp. 234–235
____ *Interactive Chalkboard CD-ROM,* Lesson 4-7
____ algebra1.com/extra_examples
____ *Guide to Daily Intervention,* pp. 14–15
____ Daily Intervention, *TWE*, p. 236
____ Study Guide and Intervention, *CRM*, pp. 249–250
____ Reading to Learn Mathematics, *CRM*, p. 253
____ *TeacherWorks CD-ROM*
____ Reading Mathematics, *SE*, p. 239

3 Practice/Apply
____ Skills Practice, *CRM*, p. 251
____ Practice, *CRM*, p. 252
____ Extra Practice, *SE*, p. 830
____ Differentiated Instruction, *TWE*, p. 235
____ *Parent and Student Study Guide Workbook,* p. 35
____ *Answer Key Transparencies,* Lesson 4-7

Assignment Guide, pp. 236–238, *SE*			
	Objective 1	Objective 2	Other
Basic	15–19 odd	21–35 odd, 39, 41	45–49, 56–80
Average	15–19 odd	21–43 odd	45–59, 56–80
Advanced	16–20 even	22–44 even	50–74 (optional: 75–80)
Reading Mathematics	1–6		
Alternate Assignment			

4 Assess
____ Open-Ended Assessment, *TWE*, p. 238
____ Enrichment, *CRM*, p. 254
____ algebra1.com/self_check_quiz

____ *Closing the Gap for Absent Students,* pp. 8–9

KEY *SE* = Student Edition *TWE* = Teacher Wraparound Edition *CRM* = Chapter Resource Masters

© Glencoe/McGraw-Hill   Glencoe Algebra 1

Lesson Planning Guide (pp. 240–245)

4-8

Teacher's Name _____ Dates _____

Grade _____ Class _____ M Tu W Th F

NCTM Standards
1, 2, 3, 6, 8, 9, 10

Recommended Pacing	
Regular Average	Day 12 of 14
Regular Advanced	Day 12 of 14
Block Average	Day 6 of 7
Block Advanced	Day 6 of 7

Objectives
___ Look for a pattern.
___ Write an equation given some of the solutions.
___ State/local objectives: _____

1 Focus
Materials/Resources Needed _____
___ *5-Minute Check Transparencies*, Lesson 4-8
___ *Mathematical Background, TWE*, p. 190D
___ *TeacherWorks CD-ROM*

2 Teach
___ In-Class Examples, *TWE*, pp. 241–242
___ *Teaching Algebra with Manipulatives*, pp. 90–92
___ *Interactive Chalkboard CD-ROM*, Lesson 4-8
___ algebra1.com/extra_examples
___ *Guide to Daily Intervention*, pp. 14–15
___ Study Guide and Intervention, *CRM*, pp. 255–256
___ Reading to Learn Mathematics, *CRM*, p. 259
___ *TeacherWorks CD-ROM*

3 Practice/Apply
___ Skills Practice, *CRM*, p. 257
___ Practice, *CRM*, p. 258
___ Extra Practice, *SE*, p. 830
___ Differentiated Instruction, *TWE*, p. 242
___ *Parent and Student Study Guide Workbook*, p. 36
___ *Answer Key Transparencies*, Lesson 4-8
___ *WebQuest and Projects Resources*, p. 30

Assignment Guide, pp. 243–245, *SE*			
	Objective 1	Objective 2	Other
Basic	13–19 odd, 27, 28	21, 23, 32	31, 33–42
Average	13–19 odd, 27, 28	21–25 odd, 32	31, 33–42
Advanced	12–18 even, 26–28	20–24 even, 29, 32	30, 31, 33–42

Alternate Assignment _____

4 Assess
___ Open-Ended Assessment, *TWE*, p. 245
___ Enrichment, *CRM*, p. 260
___ Assessment, Quiz, *CRM*, p. 276
___ algebra1.com/self_check_quiz

___ *Closing the Gap for Absent Students*, pp. 8–9

KEY *SE* = Student Edition *TWE* = Teacher Wraparound Edition *CRM* = Chapter Resource Masters

© Glencoe/McGraw-Hill *Glencoe Algebra 1*

Review and Testing (pp. 246–253)

4

Teacher's Name _____ Dates _____

Grade _____ Class _____ M Tu W Th F

Recommended Pacing	
Regular Average	Days 13 & 14 of 14
Regular Advanced	Days 13 & 14 of 14
Block Average	Day 7 of 7
Block Advanced	Day 7 of 7

Assess
___ *Parent and Student Study Guide Workbook*, p. 37
___ Vocabulary and Concept Check, *SE*, p. 246
___ Vocabulary Test, *CRM*, p. 274
___ Lesson-by-Lesson Review, *SE*, pp. 246–250
___ Practice Test, *SE*, p. 251
___ Chapter 4 Tests, *CRM*, pp. 261–272
___ Open-Ended Assessment, *CRM*, p. 273
___ Standardized Test Practice, *SE*, pp. 252–253
___ Standardized Test Practice, *CRM*, pp. 279–280
___ Cumulative Review, *CRM*, p. 278
___ *Vocabulary PuzzleMaker CD-ROM*
___ algebra1.com/vocabulary_review
___ algebra1.com/chapter_test
___ algebra1.com/standardized_test
___ *MindJogger Videoquizzes VHS*

Other Assessment Materials
- *TestCheck and Worksheet Builder CD-ROM*

KEY   SE = Student Edition   TWE = Teacher Wraparound Edition   CRM = Chapter Resource Masters

Lesson Planning Guide (pp. 256–263)

5-1

Teacher's Name _____ Dates _____
Grade _____ Class _____ M Tu W Th F

NCTM Standards
2, 3, 4, 6, 7, 8, 9, 10

Recommended Pacing	
Regular Average	Day 1 of 14
Regular Advanced	Day 1 of 13
Block Average	Day 1 of 7
Block Advanced	Day 1 of 7

Objectives
___ Find the slope of a line.
___ Use rate of change to solve problems.
___ State/local objectives: _____

1 Focus
Materials/Resources Needed _____
___ Building on Prior Knowledge, *TWE*, p. 256
___ *5-Minute Check Transparencies*, Lesson 5-1
___ Mathematical Background, *TWE*, p. 254C
___ *Prerequisite Skills Masters*, pp. 39–40, 63–64
___ *TeacherWorks CD-ROM*

2 Teach
___ In-Class Examples, *TWE*, pp. 257–258
___ *Teaching Algebra with Manipulatives*, pp. 96–97
___ *Interactive Chalkboard CD-ROM*, Lesson 5-1
___ algebra1.com/extra_examples
___ algebra1.com/usa_today
___ *Guide to Daily Intervention*, pp. 16–17
___ Daily Intervention, *TWE*, pp. 257, 259
___ Study Guide and Intervention, *CRM*, pp. 281–282
___ Reading to Learn Mathematics, *CRM*, p. 285
___ *TeacherWorks CD-ROM*
___ Reading Mathematics, *SE*, p. 263

3 Practice/Apply
___ Skills Practice, *CRM*, p. 283
___ Practice, *CRM*, p. 284
___ Extra Practice, *SE*, p. 831
___ Differentiated Instruction, *TWE*, p. 260
___ *Parent and Student Study Guide Workbook*, p. 38
___ *Answer Key Transparencies*, Lesson 5-1

Assignment Guide, pp. 259–262, *SE*			
	Objective 1	Objective 2	Other
Basic	15–29 odd, 37–43 odd, 57	50–55	58–60, 63–85
Average	15–47 odd, 57	50–55	49, 58–60, 63–85 (optional: 61, 62)
Advanced	16–38 even, 42–48 even, 57	40, 53–56	49, 58–76 (optional: 77–85)
Reading Mathematics	1–2		
Alternate Assignment			

4 Assess
___ Open-Ended Assessment, *TWE*, p. 262
___ Enrichment, *CRM*, p. 286
___ algebra1.com/self_check_quiz

___ *Closing the Gap for Absent Students*, pp. 10–11

KEY   *SE* = Student Edition   *TWE* = Teacher Wraparound Edition   *CRM* = Chapter Resource Masters

© Glencoe/McGraw-Hill   Glencoe Algebra 1

Lesson Planning Guide (pp. 264–270)

5-2

Teacher's Name _____ Dates _____
Grade _____ Class _____ M Tu W Th F

NCTM Standards
2, 4, 8, 9, 10

Recommended Pacing	
Regular Average	Day 2 of 14
Regular Advanced	Day 2 of 13
Block Average	Day 1 of 7
Block Advanced	Day 1 of 7

Objectives
____ Write and graph direct variation equations.
____ Solve problems involving direct variation.
____ State/local objectives: _____

1 Focus
Materials/Resources Needed _____

____ Building on Prior Knowledge, *TWE*, p. 264
____ *5-Minute Check Transparencies*, Lesson 5-2
____ Mathematical Background, *TWE*, p. 254C
____ *Prerequisite Skills Masters*, pp. 29–30
____ *TeacherWorks CD-ROM*

2 Teach
____ In-Class Examples, *TWE*, pp. 265–266
____ *Interactive Chalkboard CD-ROM*, Lesson 5-2
____ algebra1.com/extra_examples
____ algebra1.com/career_choices
____ *Guide to Daily Intervention*, pp. 16–17
____ Study Guide and Intervention, *CRM*, pp. 287–288
____ Reading to Learn Mathematics, *CRM*, p. 291
____ *TeacherWorks CD-ROM*
____ *Multimedia Applications Masters*

3 Practice/Apply
____ Skills Practice, *CRM*, p. 289
____ Practice, *CRM*, p. 290
____ Extra Practice, *SE*, p. 831
____ Differentiated Instruction, *TWE*, p. 266
____ *Science and Mathematics Lab Manual*, pp. 43–46
____ *Graphing Calculator and Spreadsheet Masters*, p. 54
____ *Parent and Student Study Guide Workbook*, p. 39
____ *Answer Key Transparencies*, Lesson 5-2

Assignment Guide, pp. 267–270, *SE*			
	Objective 1	Objective 2	Other
Basic	15–41 odd, 47	43, 45, 49, 51–53	56–58, 63–78
Average	15–41 odd, 47	43, 45, 49, 51, 55	54, 56–58, 63–78 (optional: 59–62)
Advanced	16–42 even, 47	44–50 even, 54, 55	56–72 (optional: 73–78)
All	Practice Quiz 1 (1–10)		
Alternate Assignment			

4 Assess
____ Practice Quiz 1, *SE*, p. 270
____ Open-Ended Assessment, *TWE*, p. 270
____ Enrichment, *CRM*, p. 292
____ Assessment, Quiz, *CRM*, p. 337
____ algebra1.com/self_check_quiz

____ Closing the Gap for Absent Students, pp. 10–11

KEY *SE* = Student Edition   *TWE* = Teacher Wraparound Edition   *CRM* = Chapter Resource Masters

© Glencoe/McGraw-Hill   Glencoe Algebra 1

Algebra Activity (p. 271)
A Preview of Lesson 5-3

Teacher's Name _____ Dates _____

Grade _____ Class _____ M Tu W Th F

NCTM Standards
2, 3, 4, 8, 9, 10

Recommended Pacing	
Regular Average	Day 3 of 14
Regular Advanced	Day 3 of 13
Block Average	Day 2 of 7
Block Advanced	Day 2 of 7

Objectives
___ Use manipulatives to investigate slope-intercept form.
___ State/local objectives: _____

Getting Started
Materials/Resources Needed plastic sandwich bag, long rubber band, tape, centimeter ruler, scissors, grid paper, washers

Teach
___ *Teaching Algebra with Manipulatives,* p. 98
___ *Glencoe Mathematics Classroom Manipulative Kit*
___ Teaching Strategy, *TWE,* p. 271

Assignment Guide, p. 271, *SE*	
All	1–8
Alternate Assignment	

Assess
___ Study Notebook, *TWE,* p. 271

KEY *SE* = Student Edition   *TWE* = Teacher Wraparound Edition   *CRM* = Chapter Resource Masters

© Glencoe/McGraw-Hill   Glencoe Algebra 1

Lesson Planning Guide (pp. 272–277)

5-3

Teacher's Name _____ Dates _____

Grade _____ Class _____ M Tu W Th F

NCTM Standards
2, 3, 4, 6, 8, 9, 10

Recommended Pacing	
Regular Average	Days 3 & 4 of 14
Regular Advanced	Days 3 & 4 of 13
Block Average	Day 2 of 7
Block Advanced	Day 2 of 7

Objectives
___ Write and graph linear equations in slope-intercept form.
___ Model real-world data with an equation in slope-intercept form.
___ State/local objectives: _____

---

1 Focus
Materials/Resources Needed _____
_____
___ *5-Minute Check Transparencies,* Lesson 5-3
___ *Mathematical Background, TWE,* p. 254C
___ *TeacherWorks CD-ROM*

2 Teach
___ In-Class Examples, *TWE,* pp. 273–274
___ *Teaching Algebra with Manipulatives,* p. 99
___ *Interactive Chalkboard CD-ROM,* Lesson 5-3
___ algebra1.com/extra_examples
___ *Guide to Daily Intervention,* pp. 16–17
___ *Study Guide and Intervention, CRM,* pp. 293–294
___ *Reading to Learn Mathematics, CRM,* p. 297
___ *TeacherWorks CD-ROM*
___ *Multimedia Applications Masters*

3 Practice/Apply
___ *Skills Practice, CRM,* p. 295
___ *Practice, CRM,* p. 296
___ *Extra Practice, SE,* p. 831
___ *Differentiated Instruction, TWE,* p. 274
___ *School-to-Career Masters,* p. 9
___ *Parent and Student Study Guide Workbook,* p. 40
___ *Real-World Transparency and Master*
___ *Answer Key Transparencies,* Lesson 5-3

Assignment Guide, pp. 275–277, *SE*			
	Objective 1	Objective 2	Other
Basic	15–39 odd	41–45 odd, 46	42, 50–52, 56–67
Average	15–39 odd	41–45 odd, 46	44, 50–52, 56–67 (optional: 53–55)
Advanced	14–38 even	45–49	40–44 even 50–64 (optional: 65–67)

Alternate Assignment _____

4 Assess
___ Open-Ended Assessment, *TWE,* p. 277
___ Enrichment, *CRM,* p. 298
___ algebra1.com/self_check_quiz

___ *Closing the Gap for Absent Students,* pp. 10–11

KEY	*SE* = Student Edition	*TWE* = Teacher Wraparound Edition	*CRM* = Chapter Resource Masters

© Glencoe/McGraw-Hill — Glencoe Algebra 1

Graphing Calculator Investigation (pp. 278–279)
A Follow-Up of Lesson 5-3

5

Teacher's Name _____ Dates _____

Grade _____ Class _____ M Tu W Th F

NCTM Standards
2, 6, 7, 8, 10

Recommended Pacing	
Regular Average	Day 4 of 14
Regular Advanced	Day 4 of 13
Block Average	Day 2 of 7
Block Advanced	Day 2 of 7

Objectives
___ Use a graphing calculator to identify families of linear graphs.
___ State/local objectives: _____

Getting Started
Materials/Resources Needed _____
_____
_____

Teach
___ *Graphing Calculator and Spreadsheet Masters*, p. 55
___ algebra1.com/other_calculator_keystrokes

Assignment Guide, p. 279, *SE*	
All	1–9
Alternate Assignment	

Assess

KEY *SE* = Student Edition   *TWE* = Teacher Wraparound Edition   *CRM* = Chapter Resource Masters

© Glencoe/McGraw-Hill   Glencoe Algebra 1

Lesson Planning Guide (pp. 280–285)

5-4

Teacher's Name _____ Dates _____

Grade _____ Class _____ M Tu W Th F

NCTM Standards
2, 6, 8, 9, 10

Recommended Pacing	
Regular Average	Days 5 & 6 of 14
Regular Advanced	Days 5 & 6 of 13
Block Average	Day 3 of 7
Block Advanced	Day 3 of 7

Objectives

___ Write an equation of a line given the slope and one point on a line.
___ Write an equation of a line given two points on the line.
___ State/local objectives: _____

1 Focus

Materials/Resources Needed _____

___ *5-Minute Check Transparencies*, Lesson 5-4
___ *Mathematical Background*, *TWE*, p. 254D
___ *TeacherWorks CD-ROM*

2 Teach

___ In-Class Examples, *TWE*, pp. 281–282
___ *Interactive Chalkboard CD-ROM*, Lesson 5-4
___ algebra1.com/extra_examples
___ *Guide to Daily Intervention*, pp. 16–17
___ Study Guide and Intervention, *CRM*, pp. 299–300
___ Reading to Learn Mathematics, *CRM*, p. 303
___ *TeacherWorks CD-ROM*

3 Practice/Apply

___ Skills Practice, *CRM*, p. 301
___ Practice, *CRM*, p. 302
___ Extra Practice, *SE*, p. 832
___ Differentiated Instruction, *TWE*, p. 282
___ *School-to-Career Masters*, p. 10
___ *Parent and Student Study Guide Workbook*, p. 41
___ *Answer Key Transparencies*, Lesson 5-4
___ *AlgePASS CD-ROM*, Lesson 11

Assignment Guide, pp. 283–285, *SE*			
	Objective 1	Objective 2	Other
Basic	11–17 odd	19–27 odd, 38	39, 44–62
Average	11–17 odd	19–33 odd, 34–38 even	39, 44–62
Advanced	12–18 even	20–32 even, 38, 41	39, 40, 42–56 (optional: 57–62)

Alternate Assignment _____

4 Assess

___ Open-Ended Assessment, *TWE*, p. 285
___ Enrichment, *CRM*, p. 304
___ Assessment, Mid-Chapter Test, *CRM*, p. 339
___ Assessment, Quiz, *CRM*, p. 337
___ algebra1.com/self_check_quiz

___ *Closing the Gap for Absent Students*, pp. 10–11

KEY *SE* = Student Edition *TWE* = Teacher Wraparound Edition *CRM* = Chapter Resource Masters

© Glencoe/McGraw-Hill Glencoe Algebra 1

Lesson Planning Guide (pp. 286–291)

5-5

Teacher's Name _____ Dates _____

Grade _____ Class _____ M Tu W Th F

NCTM Standards
2, 6, 8, 9, 10

Recommended Pacing	
Regular Average	Day 7 of 14
Regular Advanced	Days 7 & 8 of 13
Block Average	Day 4 of 7
Block Advanced	Day 4 of 7

Objectives
___ Write the equation of a line in point-slope form.
___ Write linear equations in different forms.
___ State/local objectives: _____

1 Focus
Materials/Resources Needed _____

___ *5-Minute Check Transparencies,* Lesson 5-5
___ *Mathematical Background, TWE,* p. 254D
___ *TeacherWorks CD-ROM*

2 Teach
___ *In-Class Examples, TWE,* pp. 287–288
___ *Teaching Algebra with Manipulatives,* pp. 100–101
___ *Interactive Chalkboard CD-ROM,* Lesson 5-5
___ algebra1.com/extra_examples
___ algebra1.com/data_update
___ *Guide to Daily Intervention,* pp. 16–17
___ *Daily Intervention, TWE,* p. 289
___ *Study Guide and Intervention, CRM,* pp. 305–306
___ *Reading to Learn Mathematics, CRM,* p. 309
___ *TeacherWorks CD-ROM*

3 Practice/Apply
___ *Skills Practice, CRM,* p. 307
___ *Practice, CRM,* p. 308
___ *Extra Practice, SE,* p. 832
___ *Differentiated Instruction, TWE,* p. 288
___ *Graphing Calculator and Spreadsheet Masters,* p. 31
___ *Parent and Student Study Guide Workbook,* p. 42
___ *Answer Key Transparencies,* Lesson 5-5
___ *AlgePASS CD-ROM,* Lesson 12

Assignment Guide, pp. 289–291, SE			
	Objective 1	Objective 2	Other
Basic	15–27 odd	29–53 odd	55–57, 61–67, 72–87
Average	15–27 odd	29–53 odd	58–67, 72–87 (optional: 68–71)
Advanced	16–28 even	30–54 even	58–79 (optional: 80–87)

Alternate Assignment _____

4 Assess
___ Open-Ended Assessment, *TWE,* p. 291
___ Enrichment, *CRM,* p. 310
___ algebra1.com/self_check_quiz

___ *Closing the Gap for Absent Students,* pp. 10–11

KEY *SE* = Student Edition   *TWE* = Teacher Wraparound Edition   *CRM* = Chapter Resource Masters

© Glencoe/McGraw-Hill   Glencoe Algebra 1

Lesson Planning Guide (pp. 292–297)

5-6

Teacher's Name _____ Dates _____

Grade _____ Class _____ M Tu W Th F

NCTM Standards
2, 3, 6, 7, 8, 9, 10

Recommended Pacing	
Regular Average	Days 8 & 9 of 14
Regular Advanced	Day 9 of 13
Block Average	Days 4 & 5 of 7
Block Advanced	Day 5 of 7

Objectives

___ Write an equation of the line that passes through a given point, parallel to a given line.
___ Write an equation of the line that passes through a given point, perpendicular to a given line.
___ State/local objectives: _____

1 Focus

Materials/Resources Needed _____
_____

___ Building on Prior Knowledge, *TWE*, p. 292
___ 5-Minute Check Transparencies, Lesson 5-6
___ Mathematical Background, *TWE*, p. 254D
___ *TeacherWorks CD-ROM*

2 Teach

___ In-Class Examples, *TWE*, pp. 293–294
___ *Teaching Algebra with Manipulatives*, pp. 102–104
___ *Interactive Chalkboard CD-ROM*, Lesson 5-6
___ algebra1.com/extra_examples
___ *Guide to Daily Intervention*, pp. 16–17
___ Study Guide and Intervention, *CRM*, pp. 311–312
___ *Reading to Learn Mathematics*, *CRM*, p. 315
___ *TeacherWorks CD-ROM*

3 Practice/Apply

___ Skills Practice, *CRM*, p. 313
___ Practice, *CRM*, p. 314
___ Extra Practice, *SE*, p. 832
___ Differentiated Instruction, *TWE*, p. 294
___ *Parent and Student Study Guide Workbook*, p. 43
___ *Answer Key Transparencies*, Lesson 5-6
___ *AlgePASS CD-ROM*, Lesson 13

Assignment Guide, pp. 295–297, *SE*			
	Objective 1	Objective 2	Other
Basic	13–27 odd	29–37 odd, 41	46–60
Average	13–27 odd	29–41 odd	43, 45–60
Advanced	14–26 even	30–40 even	42, 44, 46–54 (optional: 55–60)
All	Practice Quiz 2 (1–5)		
Alternate Assignment	_____		

4 Assess

___ Practice Quiz 2, *SE*, p. 297
___ Open-Ended Assessment, *TWE*, p. 297
___ Enrichment, *CRM*, p. 316

___ Assessment, Quiz, *CRM*, p. 338
___ algebra1.com/self_check_quiz

___ Closing the Gap for Absent Students, pp. 10–11

KEY *SE* = Student Edition *TWE* = Teacher Wraparound Edition *CRM* = Chapter Resource Masters

© Glencoe/McGraw-Hill   Glencoe Algebra 1

Lesson Planning Guide (pp. 298–305)

5-7

Teacher's Name _____ Dates _____

Grade _____ Class _____ M Tu W Th F

NCTM Standards
2, 5, 6, 7, 8, 9, 10

Recommended Pacing	
Regular Average	Days 10 & 11 of 14
Regular Advanced	Day 10 of 13
Block Average	Days 5 & 6 of 7
Block Advanced	Days 5 & 6 of 7

Objectives
____ Interpret points on a scatter plot.
____ Write equations for lines of fit.
____ State/local objectives: _____

---

1 Focus
Materials/Resources Needed _____

____ Building on Prior Knowledge, *TWE*, p. 298
____ 5-Minute Check Transparencies, Lesson 5-7
____ Mathematical Background, *TWE*, p. 254D
____ TeacherWorks CD-ROM

2 Teach
____ In-Class Examples, *TWE*, pp. 299–301
____ *Teaching Algebra with Manipulatives*, pp. 105–107
____ *Interactive Chalkboard CD-ROM*, Lesson 5-7
____ algebra1.com/extra_examples
____ algebra1.com/careers
____ algebra1.com/data_updates
____ *Guide to Daily Intervention*, pp. 16–17
____ Study Guide and Intervention, *CRM*, pp. 317–318
____ Reading to Learn Mathematics, *CRM*, p. 321
____ TeacherWorks CD-ROM

---

3 Practice/Apply
____ Skills Practice, *CRM*, p. 319
____ Practice, *CRM*, p. 320
____ Extra Practice, *SE*, p. 833
____ Differentiated Instruction, *TWE*, p. 304
____ *Science and Mathematics Lab Manual*, pp. 53–58
____ *Graphing Calculator and Spreadsheet Masters*, p. 32
____ *Parent and Student Study Guide Workbook*, p. 44
____ Answer Key Transparencies, Lesson 5-7
____ *WebQuest and Projects Resources*, p. 37

Assignment Guide, pp. 301–305, *SE*			
	Objective 1	Objective 2	Other
Basic	11, 13	14–17	34–39, 45–55
Average	11, 13	19–23, 25–28	18, 24, 34–39, 45–55 (optional: 40–44)
Advanced	10, 12	25–28, 30–33	24, 29, 34–55
Alternate Assignment			

---

4 Assess
____ Open-Ended Assessment, *TWE*, p. 305
____ Enrichment, *CRM*, p. 322
____ Assessment, Quiz, *CRM*, p. 338
____ algebra1.com/self_check_quiz

____ *Closing the Gap for Absent Students*, pp. 10–11

KEY   *SE* = Student Edition   *TWE* = Teacher Wraparound Edition   *CRM* = Chapter Resource Masters

© Glencoe/McGraw-Hill   Glencoe Algebra 1

# Graphing Calculator Investigation (pp. 306–307)
A Follow-Up of Lesson 5-7

5

Teacher's Name _____ Dates _____

Grade _____ Class _____ M Tu W Th F

NCTM Standards
2, 5, 7, 8, 9, 10

Recommended Pacing	
Regular Average	Day 12 of 14
Regular Advanced	Day 11 of 13
Block Average	Day 6 of 7
Block Advanced	Day 6 of 7

Objectives
___ Use a graphing calculator to find a median-fit line.
___ State/local objectives: _____

Getting Started
Materials/Resources Needed _____
_____

Teach
___ *Graphing Calculator and Spreadsheet Masters*, p. 56
___ algebra1.com/other_calculator_keystrokes

Assignment Guide, p. 307, *SE*	
All	1–5
Alternate Assignment	

Assess

KEY *SE* = Student Edition   *TWE* = Teacher Wraparound Edition   *CRM* = Chapter Resource Masters

© Glencoe/McGraw-Hill   Glencoe Algebra 1

5 Review and Testing (pp. 308–315)

Teacher's Name _____ Dates _____

Grade _____ Class _____ M Tu W Th F

Recommended Pacing	
Regular Average	Days 13 & 14 of 14
Regular Advanced	Days 12 & 13 of 13
Block Average	Days 6 & 7 of 7
Block Advanced	Days 6 & 7 of 7

Assess

____ *Parent and Student Study Guide Workbook*, p. 45
____ Vocabulary and Concept Check, *SE*, p. 308
____ Vocabulary Test, *CRM*, p. 336
____ Lesson-by-Lesson Review, *SE*, pp. 308–312
____ Practice Test, *SE*, p. 313
____ Chapter 5 Tests, *CRM*, pp. 323–334
____ Open-Ended Assessment, *CRM*, p. 335
____ Standardized Test Practice, *SE*, pp. 314–315
____ Standardized Test Practice, *CRM*, pp. 341–342
____ Cumulative Review, *CRM*, p. 340
____ *Vocabulary PuzzleMaker CD-ROM*
____ algebra1.com/vocabulary_review
____ algebra1.com/chapter_test
____ algebra1.com/standardized_test
____ *MindJogger Videoquizzes VHS*

Other Assessment Materials

- *TestCheck and Worksheet Builder CD-ROM*

KEY   *SE* = Student Edition   *TWE* = Teacher Wraparound Edition   *CRM* = Chapter Resource Masters

© Glencoe/McGraw-Hill   Glencoe Algebra 1

Lesson Planning Guide (pp. 318–323)

6-1

Teacher's Name _____ Dates _____

Grade _____ Class _____ M Tu W Th F

NCTM Standards
2, 6, 8, 9, 10

Recommended Pacing	
Regular Average	Day 1 of 12
Regular Advanced	Day 1 of 11
Block Average	Day 1 of 7
Block Advanced	Day 1 of 6

Objectives
___ Solve linear inequalities by using addition.
___ Solve linear inequalities by using subtraction.
___ State/local objectives: _____

1 Focus
Materials/Resources Needed _____

___ Building on Prior Knowledge, *TWE*, p. 318
___ *5-Minute Check Transparencies*, Lesson 6-1
___ Mathematical Background, *TWE*, p. 316C
___ *TeacherWorks CD-ROM*

2 Teach
___ In-Class Examples, *TWE*, pp. 319–320
___ *Teaching Algebra with Manipulatives*, pp. 111–114
___ *Interactive Chalkboard CD-ROM*, Lesson 6-1
___ algebra1.com/extra_examples
___ *Guide to Daily Intervention*, pp. 18–19
___ Daily Intervention, *TWE*, p. 321
___ Study Guide and Intervention, *CRM*, pp. 343–344
___ Reading to Learn Mathematics, *CRM*, p. 347
___ *TeacherWorks CD-ROM*

3 Practice/Apply
___ Skills Practice, *CRM*, p. 345
___ Practice, *CRM*, p. 346
___ Extra Practice, *SE*, p. 833
___ Differentiated Instruction, *TWE*, p. 320
___ *School-to-Career Masters*, p. 11
___ *Parent and Student Study Guide Workbook*, p. 46
___ *Answer Key Transparencies*, Lesson 6-1
___ *WebQuest and Projects Resources*, p. 37

Assignment Guide, pp. 321–323, *SE*			
	Objective 1	Objective 2	Other
Basic	15–31 odd, 41–51 odd	15–31 odd, 41–51 odd	53, 56–76
Average	15–51 odd	15–51 odd	53, 56–76
Advanced	14–50 even	14–50 even	52–68 (optional: 69–76)

Alternate Assignment _____

4 Assess
___ Open-Ended Assessment, *TWE*, p. 323
___ Enrichment, *CRM*, p. 348
___ algebra1.com/self_check_quiz

___ Closing the Gap for Absent Students, pp. 12–13

KEY *SE* = Student Edition   *TWE* = Teacher Wraparound Edition   *CRM* = Chapter Resource Masters

Algebra Activity (p. 324)
A Preview of Lesson 6-2

6

Teacher's Name _____ Dates _____

Grade _____ Class _____ M Tu W Th F

NCTM Standards
2, 6, 8, 9, 10

Recommended Pacing	
Regular Average	Day 2 of 12
Regular Advanced	Day 2 of 11
Block Average	Day 1 of 7
Block Advanced	Day 1 of 6

Objectives
___ Use algebra tiles to solve inequalities.
___ State/local objectives: _____

Getting Started
Materials/Resources Needed algebra tiles, equation mat, self-adhesive notes
_____
_____
_____
_____

Teach
___ *Teaching Algebra with Manipulatives*, p. 115
___ *Glencoe Mathematics Classroom Manipulative Kit*
___ Teaching Strategy, *TWE*, p. 324

Assignment Guide, p. 324, *SE*	
All	1–7
Alternate Assignment	

Assess
___ Study Notebook, *TWE*, p. 324

KEY   *SE* = Student Edition   *TWE* = Teacher Wraparound Edition   *CRM* = Chapter Resource Masters

© Glencoe/McGraw-Hill   *Glencoe Algebra 1*

6-2 Lesson Planning Guide (pp. 325–331)

Teacher's Name _____ Dates _____

Grade _____ Class _____ M Tu W Th F

NCTM Standards
2, 6, 8, 9, 10

Recommended Pacing	
Regular Average	Day 3 of 12
Regular Advanced	Day 2 of 11
Block Average	Day 2 of 7
Block Advanced	Day 1 of 6

Objectives

___ Solve linear inequalities by using multiplication.
___ Solve linear inequalities by using division.
___ State/local objectives: _____

1 Focus

Materials/Resources Needed _____

___ Building on Prior Knowledge, *TWE*, p. 325
___ *5-Minute Check Transparencies*, Lesson 6-2
___ Mathematical Background, *TWE*, p. 316C
___ *TeacherWorks CD-ROM*

2 Teach

___ In-Class Examples, *TWE*, pp. 326–328
___ *Interactive Chalkboard CD-ROM*, Lesson 6-2
___ algebra1.com/extra_examples
___ *Guide to Daily Intervention*, pp. 18–19
___ Daily Intervention, *TWE*, p. 329
___ Study Guide and Intervention, *CRM*, pp. 349–350
___ Reading to Learn Mathematics, *CRM*, p. 353
___ *TeacherWorks CD-ROM*

3 Practice/Apply

___ Skills Practice, *CRM*, p. 351
___ Practice, *CRM*, p. 352
___ Extra Practice, *SE*, p. 833
___ Differentiated Instruction, *TWE*, p. 327
___ *School-to-Career Masters*, p. 12
___ *Parent and Student Study Guide Workbook*, p. 47
___ *Answer Key Transparencies*, Lesson 6-2
___ *WebQuest and Projects Resources*, p. 38

Assignment Guide, pp. 328–331, *SE*			
	Objective 1	Objective 2	Other
Basic	13–29 odd, 35, 39, 41 45–49 odd	13–29 odd, 35, 39, 41 45–49 odd	52, 55–78
Average	13–49 odd	13–49 odd	51–53, 55–78
Advanced	14–50 even	14–50 even	52, 54–72 (optional: 73–78)

Alternate Assignment _____

4 Assess

___ Practice Quiz 1, *SE*, p. 331
___ Open-Ended Assessment, *TWE*, p. 331
___ Enrichment, *CRM*, p. 354
___ Assessment, Quiz, *CRM*, p. 393
___ algebra1.com/self_check_quiz

___ *Closing the Gap for Absent Students*, pp. 12–13

KEY *SE* = Student Edition *TWE* = Teacher Wraparound Edition *CRM* = Chapter Resource Masters

© Glencoe/McGraw-Hill *Glencoe Algebra 1*

Lesson Planning Guide (pp. 332–338)

6-3

Teacher's Name _____ Dates _____

Grade _____ Class _____ M Tu W Th F

NCTM Standards
2, 6, 8, 9, 10

Recommended Pacing	
Regular Average	Day 4 of 12
Regular Advanced	Day 3 of 12
Block Average	Days 2 & 3 of 7
Block Advanced	Day 2 of 6

Objectives
___ Solve linear inequalities involving more than one operation.
___ Solve linear inequalities involving the Distributive Property.
___ State/local objectives: _____

1 Focus
Materials/Resources Needed _____
___ Building on Prior Knowledge, *TWE*, p. 332
___ *5-Minute Check Transparencies*, Lesson 6-3
___ Mathematical Background, *TWE*, p. 316C
___ *TeacherWorks CD-ROM*

2 Teach
___ In-Class Examples, *TWE*, pp. 333–334
___ *Interactive Chalkboard CD-ROM*, Lesson 6-3
___ algebra1.com/extra_examples
___ *Guide to Daily Intervention*, pp. 18–19
___ Daily Intervention, *TWE*, p. 334
___ Study Guide and Intervention, *CRM*, pp. 355–356
___ Reading to Learn Mathematics, *CRM*, p. 359
___ *TeacherWorks CD-ROM*
___ *Multimedia Applications Masters*
___ Reading Mathematics, *SE*, p. 338

3 Practice/Apply
___ Skills Practice, *CRM*, p. 357
___ Practice, *CRM*, p. 358
___ Extra Practice, *SE*, p. 834
___ Differentiated Instruction, *TWE*, p. 335
___ *Parent and Student Study Guide Workbook*, p. 48
___ *Answer Key Transparencies*, Lesson 6-3
___ *AlgePASS CD-ROM*, Lesson 14

Assignment Guide, pp. 334–337, *SE*			
	Objective 1	Objective 2	Other
Basic	11, 15–25 odd, 35, 37, 39–42	13	46, 53–82
Average	11, 15–25 odd, 31, 35, 37, 45	13, 27, 29, 33, 43, 44	46, 53–82
Advanced	12, 16–26 even, 36, 48–52	14, 28–34, 38, 47	46, 53–73 (optional: 74–82)
Reading Mathematics	1–12		
Alternate Assignment			

4 Assess
___ Open-Ended Assessment, *TWE*, p. 337
___ Enrichment, *CRM*, p. 360
___ Assessment, Mid-Chapter Test, *CRM*, p. 395
___ Assessment, Quiz, *CRM*, p. 393
___ algebra1.com/self_check_quiz

___ *Closing the Gap for Absent Students*, pp. 12–13

KEY *SE* = Student Edition   *TWE* = Teacher Wraparound Edition   *CRM* = Chapter Resource Masters

© Glencoe/McGraw-Hill   Glencoe Algebra 1

Lesson Planning Guide (pp. 339–344)

6-4

Teacher's Name _____ Dates _____

Grade _____ Class _____ M Tu W Th F

NCTM Standards
2, 4, 6, 8, 9, 10

Recommended Pacing	
Regular Average	Days 5 & 6 of 12
Regular Advanced	Days 4 & 5 of 11
Block Average	Days 3 & 4 of 7
Block Advanced	Days 2 & 3 of 6

Objectives

____ Solve compound inequalities containing the word *and* and graph their solution sets.

____ Solve compound inequalities containing the word *or* and graph their solution sets.

____ State/local objectives: _____

1 Focus

Materials/Resources Needed _____

____ Building on Prior Knowledge, *TWE*, p. 339
____ *5-Minute Check Transparencies*, Lesson 6-4
____ Mathematical Background, *TWE*, p. 316D
____ *TeacherWorks CD-ROM*

2 Teach

____ In-Class Examples, *TWE*, pp. 340–341
____ *Teaching Algebra with Manipulatives*, pp. 116–118
____ *Interactive Chalkboard CD-ROM*, Lesson 6-4
____ algebra1.com/extra_examples
____ algebra1.com/careers
____ *Guide to Daily Intervention*, pp. 18–19
____ Study Guide and Intervention, *CRM*, pp. 361–362
____ Reading to Learn Mathematics, *CRM*, p. 365
____ *TeacherWorks CD-ROM*

3 Practice/Apply

____ Skills Practice, *CRM*, p. 363
____ Practice, *CRM*, p. 364
____ Extra Practice, *SE*, p. 834
____ Differentiated Instruction, *TWE*, p. 340
____ Parent and Student Study Guide Workbook, p. 49
____ Real-World Transparency and Master
____ Answer Key Transparencies, Lesson 6-4
____ AlgePASS CD-ROM, Lesson 15

Assignment Guide, pp. 341–344, *SE*			
	Objective 1	Objective 2	Other
Basic	15, 19, 21, 29, 33, 35, 43, 46–48	17, 23, 27, 31, 37, 45	49, 53–80
Average	15, 19, 21, 29, 33, 35, 39, 43, 47	17, 23–27 odd, 31, 37, 41, 45	49, 53–80
Advanced	14, 18, 20, 26, 28, 32, 34, 38, 42, 46, 48	16, 22, 24, 30, 36, 40, 44	49–72 (optional: 73–80)
All	Practice Quiz 2 (1–10)		
Alternate Assignment			

4 Assess

____ Practice Quiz 2, *SE*, p. 344
____ Open-Ended Assessment, *TWE*, p. 344
____ Enrichment, *CRM*, p. 366
____ algebra1.com/self_check_quiz

____ *Closing the Gap for Absent Students*, pp. 12–13

KEY *SE* = Student Edition   *TWE* = Teacher Wraparound Edition   *CRM* = Chapter Resource Masters

© Glencoe/McGraw-Hill   Glencoe Algebra 1

Lesson Planning Guide (pp. 345–351)

6-5

Teacher's Name _____ Dates _____

Grade _____ Class _____ M Tu W Th F

NCTM Standards
2, 4, 6, 8, 9, 10

Recommended Pacing	
Regular Average	Days 7 & 8 of 12
Regular Advanced	Days 6 & 7 of 11
Block Average	Days 4 & 5 of 7
Block Advanced	Days 3 & 4 of 6

Objectives
___ Solve absolute value equations.
___ Solve absolute value inequalities.
___ State/local objectives: _____

1 Focus
Materials/Resources Needed _____

___ Building on Prior Knowledge, *TWE*, p. 347
___ *5-Minute Check Transparencies,* Lesson 6-5
___ Mathematical Background, *TWE*, p. 316D
___ *Prerequisite Skills Masters,* pp. 79–80, 83–84
___ *TeacherWorks CD-ROM*

2 Teach
___ In-Class Examples, *TWE*, pp. 346–348
___ *Teaching Algebra with Manipulatives,* p. 119
___ *Interactive Chalkboard CD-ROM,* Lesson 6-5
___ algebra1.com/extra_examples
___ *Guide to Daily Intervention,* pp. 18–19
___ Daily Intervention, *TWE*, p. 348
___ Study Guide and Intervention, *CRM,* pp. 367–368
___ Reading to Learn Mathematics, *CRM,* p. 371
___ *TeacherWorks CD-ROM*

3 Practice/Apply
___ Skills Practice, *CRM*, p. 369
___ Practice, *CRM*, p. 370
___ Extra Practice, *SE*, p. 834
___ Differentiated Instruction, *TWE*, p. 346
___ *Graphing Calculator and Spreadsheet Masters,* p. 33
___ *Parent and Student Study Guide Workbook,* p. 50
___ *Answer Key Transparencies,* Lesson 6-5
___ *WebQuest and Projects Resources,* p. 38

Assignment Guide, pp. 348–351, *SE*			
	Objective 1	Objective 2	Other
Basic	19, 25, 27, 41, 55, 57	15, 17, 21, 23, 29–35 odd, 43–45 odd, 46, 47, 49, 51–53, 58	56, 59–79
Average	19, 25, 27, 41, 55, 57	15, 17, 21, 23, 29–37 odd, 45–47, 51–53, 58	39, 43, 56, 59–79
Advanced	16, 24, 26 38, 40, 55, 57	18, 20, 22 28–36 even, 42, 48–54 even, 58	44, 46, 56, 59–73 (optional: 74–79)
Alternate Assignment			

4 Assess
___ Open-Ended Assessment, *TWE*, p. 351
___ Enrichment, *CRM*, p. 372
___ Assessment, Quiz, *CRM*, p. 394
___ algebra1.com/self_check_quiz

___ Closing the Gap for Absent Students, pp. 12–13

KEY *SE* = Student Edition *TWE* = Teacher Wraparound Edition *CRM* = Chapter Resource Masters

© Glencoe/McGraw-Hill   Glencoe Algebra 1

Lesson Planning Guide (pp. 352–357)

6-6

Teacher's Name _____ Dates _____

Grade _____ Class _____ M Tu W Th F

NCTM Standards
2, 6, 8, 9, 10

Recommended Pacing	
Regular Average	Day 9 of 12
Regular Advanced	Day 8 of 11
Block Average	Day 5 of 7
Block Advanced	Day 4 of 6

Objectives
___ Graph inequalities on the coordinate plane.
___ Solve real-world problems involving linear inequalities.
___ State/local objectives: _____

1 Focus
Materials/Resources Needed _____

___ Building on Prior Knowledge, *TWE*, p. 352
___ *5-Minute Check Transparencies,* Lesson 6-6
___ Mathematical Background, *TWE*, p. 316D
___ *TeacherWorks CD-ROM*

2 Teach
___ In-Class Examples, *TWE*, pp. 353–354
___ *Teaching Algebra with Manipulatives,* pp. 120–121
___ *Interactive Chalkboard CD-ROM,* Lesson 6-6
___ algebra1.com/extra_examples
___ algebra1.com/data_update
___ *Guide to Daily Intervention,* pp. 18–19
___ Study Guide and Intervention, *CRM,* pp. 373–374
___ Reading to Learn Mathematics, *CRM,* p. 377
___ *TeacherWorks CD-ROM*

3 Practice/Apply
___ Skills Practice, *CRM,* p. 375
___ Practice, *CRM,* p. 376
___ Extra Practice, *SE,* p. 835
___ Differentiated Instruction, *TWE,* p. 353
___ *Graphing Calculator and Spreadsheet Masters,* p. 34
___ *Parent and Student Study Guide Workbook,* p. 51
___ *Answer Key Transparencies,* Lesson 6-6
___ algebra1.com/webquest

Assignment Guide, pp. 355–357, *SE*			
	Objective 1	Objective 2	Other
Basic	13–17 odd, 21–35 odd	38, 39	45–63
Average	13–37 odd	38–41	45–63
Advanced	12–36 even	40–44	45–63
Alternate Assignment			

4 Assess
___ Open-Ended Assessment, *TWE,* p. 357
___ Enrichment, *CRM,* p. 378
___ Assessment, Quiz, *CRM,* p. 394
___ algebra1.com/self_check_quiz

___ *Closing the Gap for Absent Students,* pp. 12–13

KEY *SE* = Student Edition   *TWE* = Teacher Wraparound Edition   *CRM* = Chapter Resource Masters

6 Graphing Calculator Investigation (p. 358)
A Follow-Up of Lesson 6-6

Teacher's Name _____ Dates _____

Grade _____ Class _____ M Tu W Th F

NCTM Standards
2, 6, 8, 9, 10

Recommended Pacing	
Regular Average	Day 10 of 12
Regular Advanced	Day 9 of 11
Block Average	Day 6 of 7
Block Advanced	Day 5 of 6

Objectives
___ Use a graphing calculator to investigate graphs of inequalities.
___ State/local objectives: _____

Getting Started
Materials/Resources Needed _____
_____

Teach
___ *Graphing Calculator and Spreadsheet Masters*, p. 57
___ algebra1.com/other_calculator_keystrokes

Assignment Guide, p. 358, *SE*	
All	1–3
Alternate Assignment	

Assess

KEY *SE* = Student Edition *TWE* = Teacher Wraparound Edition *CRM* = Chapter Resource Masters

© Glencoe/McGraw-Hill Glencoe Algebra 1

Review and Testing (pp. 359–365)

6

Teacher's Name _____ Dates _____

Grade _____ Class _____ M Tu W Th F

Recommended Pacing	
Regular Average	Days 11 & 12 of 12
Regular Advanced	Days 10 & 11 of 11
Block Average	Days 6 & 7 of 7
Block Advanced	Day 6 of 6

Assess

___ *Parent and Student Study Guide Workbook,* p. 52
___ Vocabulary and Concept Check, *SE,* p. 359
___ Vocabulary Test, *CRM,* p. 392
___ Lesson-by-Lesson Review, *SE,* pp. 359–362
___ Practice Test, *SE,* p. 363
___ Chapter 6 Tests, *CRM,* pp. 379–390
___ Open-Ended Assessment, *CRM,* p. 391
___ Standardized Test Practice, *SE,* pp. 364–365
___ Standardized Test Practice, *CRM,* pp. 397–398
___ Cumulative Review, *CRM,* p. 396
___ *Vocabulary PuzzleMaker CD-ROM*
___ algebra1.com/vocabulary_review
___ algebra1.com/chapter_test
___ algebra1.com/standardized_test
___ *MindJogger Videoquizzes VHS*
___ First Semester Test, *CRM,* pp. 399–402

Other Assessment Materials

- *TestCheck and Worksheet Builder CD-ROM*

KEY *SE* = Student Edition   *TWE* = Teacher Wraparound Edition   *CRM* = Chapter Resource Masters

7 Spreadsheet Investigation (p. 368)
A Preview of Lesson 7-1

Teacher's Name _____ Dates _____

Grade _____ Class _____ M Tu W Th F

NCTM Standards
1, 2, 6, 9, 10

Recommended Pacing	
Regular Average	Day 1 of 11
Regular Advanced	Day 1 of 11
Block Average	Day 1 of 6
Block Advanced	Day 1 of 6

Objectives
___ Use a spreadsheet to investigate when two quantities will be equal.
___ State/local objectives: _____

Getting Started
Materials/Resources Needed _____

Teach

Assignment Guide, p. 368, SE	
All	1–5
Alternate Assignment	

Assess

KEY   *SE* = Student Edition   *TWE* = Teacher Wraparound Edition   *CRM* = Chapter Resource Masters

© Glencoe/McGraw-Hill   Glencoe Algebra 1

Lesson Planning Guide (pp. 369–374)

7-1

Teacher's Name _____ Dates _____

Grade _____ Class _____ M Tu W Th F

NCTM Standards
2, 6, 8, 9, 10

Recommended Pacing	
Regular Average	Days 1 & 2 of 11
Regular Advanced	Days 1 & 2 of 11
Block Average	Days 1 & 2 of 6
Block Advanced	Days 1 & 2 of 6

Objectives

___ Determine whether a system of linear equations has 0, 1, or infinitely many solutions.
___ Solve systems of equations by graphing.
___ State/local objectives: _____

1 Focus

Materials/Resources Needed _____
_____

___ Building on Prior Knowledge, *TWE*, p. 370
___ *5-Minute Check Transparencies*, Lesson 7-1
___ Mathematical Background, *TWE*, p. 366C
___ *TeacherWorks CD-ROM*

2 Teach

___ In-Class Examples, *TWE*, pp. 370–371
___ *Teaching Algebra with Manipulatives*, p. 124
___ *Interactive Chalkboard CD-ROM*, Lesson 7-1
___ algebra1.com/extra_examples
___ *Guide to Daily Intervention*, pp. 20–21
___ Daily Intervention, *TWE*, p. 372
___ Study Guide and Intervention, *CRM*, pp. 403–404
___ Reading to Learn Mathematics, *CRM*, p. 407
___ *TeacherWorks CD-ROM*

3 Practice/Apply

___ Skills Practice, *CRM*, p. 405
___ Practice, *CRM*, p. 406
___ Extra Practice, *SE*, p. 835
___ Differentiated Instruction, *TWE*, p. 370
___ School-to-Career Masters, p. 13
___ Graphing Calculator and Spreadsheet Masters, p. 58
___ Parent and Student Study Guide Workbook, p. 53
___ Real-World Transparency and Master
___ Answer Key Transparencies, Lesson 7-1
___ WebQuest and Projects Resources, p. 45
___ algebra1.com/webquest

Assignment Guide, pp. 372–374, *SE*			
	Objective 1	Objective 2	Other
Basic	15–21 odd	23–41 odd, 42–47	55–68
Average	15–21 odd	23–41 odd, 42–47	55–68
Advanced	16–22 even	24–40 even, 48–54	56–64 (optional: 65–68)

Alternate Assignment _____

4 Assess

___ Open-Ended Assessment, *TWE*, p. 374
___ Enrichment, *CRM*, p. 408
___ algebra1.com/self_check_quiz

___ Closing the Gap for Absent Students, pp. 14–15

KEY   *SE* = Student Edition   *TWE* = Teacher Wraparound Edition   *CRM* = Chapter Resource Masters

© Glencoe/McGraw-Hill   Glencoe Algebra 1

Graphing Calculator Investigation (p. 375)
A Follow-Up of Lesson 7-1

Teacher's Name _____ Dates _____

Grade _____ Class _____ M Tu W Th F

NCTM Standards
2, 6

Recommended Pacing	
Regular Average	Day 2 of 11
Regular Advanced	Day 2 of 11
Block Average	Day 2 of 6
Block Advanced	Day 2 of 6

Objectives
___ Using a graphing calculator to solve a system of equations.
___ State/local objectives: _____

Getting Started
Materials/Resources Needed _____

Teach
___ algebra1.com/other_calculator_keystrokes

Assignment Guide, p. 375, SE	
All	1–10
Alternate Assignment	

Assess

KEY *SE* = Student Edition *TWE* = Teacher Wraparound Edition *CRM* = Chapter Resource Masters

Lesson Planning Guide (pp. 376–381)

7-2

Teacher's Name _____ Dates _____

Grade _____ Class _____ M Tu W Th F

NCTM Standards
2, 6, 8, 9, 10

Recommended Pacing	
Regular Average	Days 3 & 4 of 11
Regular Advanced	Days 3 & 4 of 11
Block Average	Days 2 & 3 of 6
Block Advanced	Days 2 & 3 of 6

Objectives
___ Solve systems of equations by using substitution.
___ Solve real-world problems involving systems of equations.
___ State/local objectives: _____

1 Focus
Materials/Resources Needed _____

___ *5-Minute Check Transparencies*, Lesson 7-2
___ *Mathematical Background*, *TWE*, p. 366C
___ *Prerequisite Skills Masters*, pp. 27–28
___ *TeacherWorks CD-ROM*

2 Teach
___ In-Class Examples, *TWE*, pp. 377–378
___ *Teaching Algebra with Manipulatives*, p. 125
___ *Interactive Chalkboard CD-ROM*, Lesson 7-2
___ algebra1.com/extra_examples
___ *Guide to Daily Intervention*, pp. 20–21
___ *Study Guide and Intervention*, *CRM*, pp. 409–410
___ *Reading to Learn Mathematics*, *CRM*, p. 413
___ *TeacherWorks CD-ROM*

3 Practice/Apply
___ *Skills Practice*, *CRM*, p. 411
___ *Practice*, *CRM*, p. 412
___ *Extra Practice*, *SE*, p. 835
___ *Differentiated Instruction*, *TWE*, p. 378
___ *Science and Mathematics Lab Manual*, pp. 65–68
___ *Parent and Student Study Guide Workbook*, p. 54
___ *Answer Key Transparencies*, Lesson 7-2
___ *AlgePASS CD-ROM*, Lesson 16

Assignment Guide, pp. 379–381, *SE*			
	Objective 1	Objective 2	Other
Basic	11–27 odd	29–35 odd	34, 39–53
Average	11–27 odd	29–35 odd	34, 37, 39–53
Advanced	12–28 even	30, 32, 36	37–49 (optional: 50–53)
All	Practice Quiz 1 (1–5)		
Alternate Assignment			

4 Assess
___ Practice Quiz 1, *SE*, p. 381
___ Open-Ended Assessment, *TWE*, p. 381
___ Enrichment, *CRM*, p. 414

___ Assessment, Quiz, *CRM*, p. 447
___ algebra1.com/self_check_quiz

___ *Closing the Gap for Absent Students*, pp. 14–15

KEY   *SE* = Student Edition   *TWE* = Teacher Wraparound Edition   *CRM* = Chapter Resource Masters

© Glencoe/McGraw-Hill   *Glencoe Algebra 1*

Lesson Planning Guide (pp. 382–386)

7-3

Teacher's Name _____ Dates _____

Grade _____ Class _____ M Tu W Th F

NCTM Standards
2, 6, 8, 9, 10

Recommended Pacing	
Regular Average	Days 5 & 6 of 11
Regular Advanced	Days 5 & 6 of 11
Block Average	Days 3 & 4 of 6
Block Advanced	Days 3 & 4 of 6

Objectives
____ Solve systems of equations by using elimination with addition.
____ Solve systems of equations by using elimination with subtraction.
____ State/local objectives: _____

1 Focus
Materials/Resources Needed _____

____ Building on Prior Knowledge, *TWE*, p. 383
____ *5-Minute Check Transparencies*, Lesson 7-3
____ Mathematical Background, *TWE*, p. 366D
____ *TeacherWorks CD-ROM*

2 Teach
____ In-Class Examples, *TWE*, pp. 383–384
____ *Teaching Algebra with Manipulatives*, p. 126
____ *Interactive Chalkboard CD-ROM*, Lesson 7-3
____ algebra1.com/extra_examples
____ *Guide to Daily Intervention*, pp. 20–21
____ Daily Intervention, *TWE*, p. 384
____ Study Guide and Intervention, *CRM*, pp. 415–416
____ Reading to Learn Mathematics, *CRM*, p. 419
____ *TeacherWorks CD-ROM*

3 Practice/Apply
____ Skills Practice, *CRM*, p. 417
____ Practice, *CRM*, p. 418
____ Extra Practice, *SE*, p. 836
____ Differentiated Instruction, *TWE*, p. 383
____ *Graphing Calculator and Spreadsheet Masters*, p. 36
____ *Parent and Student Study Guide Workbook*, p. 55
____ *Answer Key Transparencies*, Lesson 7-3

Assignment Guide, pp. 384–386, *SE*			
	Objective 1	Objective 2	Other
Basic	13, 15, 21, 31	17, 19, 23, 25, 29, 33, 35	40–54
Average	13, 15, 21, 31	17, 19, 23–29, 33, 35	40–54
Advanced	12, 14, 20, 22, 26–36 even	16, 18, 24	37–50 (optional: 51–54)

Alternate Assignment _____

4 Assess
____ Open-Ended Assessment, *TWE*, p. 386
____ Enrichment, *CRM*, p. 420
____ Assessment, Mid-Chapter Test, *CRM*, p. 449
____ Assessment, Quiz, *CRM*, p. 447
____ algebra1.com/self_check_quiz

____ Closing the Gap for Absent Students, pp. 14–15

KEY   *SE* = Student Edition   *TWE* = Teacher Wraparound Edition   *CRM* = Chapter Resource Masters

© Glencoe/McGraw-Hill   Glencoe Algebra 1

Lesson Planning Guide (pp. 387–393)

7-4

Teacher's Name _____ Dates _____

Grade _____ Class _____ M Tu W Th F

NCTM Standards
2, 6, 8, 9, 10

Recommended Pacing	
Regular Average	Days 7 & 8 of 11
Regular Advanced	Days 7 & 8 of 11
Block Average	Days 4 & 5 of 6
Block Advanced	Days 4 & 5 of 6

Objectives
____ Solve systems of equations by using elimination with multiplication.
____ Determine the best method for solving systems of equations.
____ State/local objectives: _____

1 Focus
Materials/Resources Needed _____
_____
____ *5-Minute Check Transparencies*, Lesson 7-4
____ Mathematical Background, *TWE*, p. 366D
____ *TeacherWorks CD-ROM*

2 Teach
____ In-Class Examples, *TWE*, pp. 388–390
____ *Interactive Chalkboard CD-ROM*, Lesson 7-4
____ algebra1.com/extra_examples
____ algebra1.com/data_update
____ *Guide to Daily Intervention*, pp. 20–21
____ Study Guide and Intervention, *CRM*, pp. 421–422
____ Reading to Learn Mathematics, *CRM*, p. 425
____ *TeacherWorks CD-ROM*
____ Reading Mathematics, *SE*, p. 393

3 Practice/Apply
____ Skills Practice, *CRM*, p. 423
____ Practice, *CRM*, p. 424
____ Extra Practice, *SE*, p. 836
____ Differentiated Instruction, *TWE*, p. 389
____ School-to-Career Masters, p. 14
____ Graphing Calculator and Spreadsheet Masters, p. 59
____ *Parent and Student Study Guide Workbook*, p. 56
____ Answer Key Transparencies, Lesson 7-4

Assignment Guide, pp. 390–392, *SE*			
	Objective 1	Objective 2	Other
Basic	13–25 odd	27–39 odd	40, 44–57
Average	13–25 odd	27–43 odd	40, 44–57
Advanced	14–26 even	28–42 even	43–53 (optional: 54–57)
All	Practice Quiz 2 (1–5)		
Reading Mathematics	1–5		
Alternate Assignment			

4 Assess
____ Practice Quiz 2, *SE*, p. 392
____ Open-Ended Assessment, *TWE*, p. 392
____ Enrichment, *CRM*, p. 426
____ Assessment, Quiz, *CRM*, p. 448
____ algebra1.com/self_check_quiz

____ Closing the Gap for Absent Students, pp. 14–15

KEY *SE* = Student Edition   *TWE* = Teacher Wraparound Edition   *CRM* = Chapter Resource Masters

Lesson Planning Guide (pp. 394–398)

7-5

Teacher's Name _____ Dates _____

Grade _____ Class _____ M Tu W Th F

NCTM Standards
2, 6, 8, 9, 10

Recommended Pacing	
Regular Average	Day 9 of 11
Regular Advanced	Day 9 of 11
Block Average	Day 5 of 6
Block Advanced	Day 5 of 6

Objectives
___ Solve systems of inequalities by graphing.
___ Solve real-world problems involving systems of inequalities.
___ State/local objectives: _____

1 Focus
Materials/Resources Needed _____
_____
___ Building on Prior Knowledge, *TWE*, p. 394
___ 5-Minute Check Transparencies, Lesson 7-5
___ Mathematical Background, *TWE*, p. 366D
___ TeacherWorks CD-ROM

2 Teach
___ In-Class Examples, *TWE*, pp. 395–396
___ *Teaching Algebra with Manipulatives*, pp. 127–128
___ *Interactive Chalkboard CD-ROM*, Lesson 7-5
___ algebra1.com/extra_examples
___ algebra1.com/careers
___ *Guide to Daily Intervention*, pp. 20–21
___ Daily Intervention, *TWE*, p. 396
___ Study Guide and Intervention, *CRM*, pp. 427–428
___ Reading to Learn Mathematics, *CRM*, p. 431
___ TeacherWorks CD-ROM
___ *Multimedia Applications Masters*

3 Practice/Apply
___ Skills Practice, *CRM*, p. 429
___ Practice, *CRM*, p. 430
___ Extra Practice, *SE*, p. 836
___ Differentiated Instruction, *TWE*, p. 396
___ *Graphing Calculator and Spreadsheet Masters*, pp. 35, 60
___ *Parent and Student Study Guide Workbook*, p. 57
___ Answer Key Transparencies, Lesson 7-5
___ AlgePASS CD-ROM, Lesson 17
___ algebra1.com/webquest

Assignment Guide, pp. 396–398, *SE*			
	Objective 1	Objective 2	Other
Basic	13–23 odd	29–31, 35	32, 39–49
Average	13–27 odd	29–31, 35	32, 39–49 (optional: 36–38)
Advanced	12–28 even	33–35	32, 36–49
Alternate Assignment			

4 Assess
___ Open-Ended Assessment, *TWE*, p. 398
___ Enrichment, *CRM*, p. 432
___ Assessment, Quiz, *CRM*, p. 448
___ algebra1.com/self_check_quiz

___ *Closing the Gap for Absent Students*, pp. 14–15

KEY   *SE* = Student Edition   *TWE* = Teacher Wraparound Edition   *CRM* = Chapter Resource Masters

© Glencoe/McGraw-Hill   *Glencoe Algebra 1*

Review and Testing (pp. 399–405)

7

Teacher's Name _____ Dates _____

Grade _____ Class _____ M Tu W Th F

Recommended Pacing	
Regular Average	Days 10 & 11 of 11
Regular Advanced	Days 10 & 11 of 11
Block Average	Day 6 of 6
Block Advanced	Day 6 of 6

Assess

____ *Parent and Student Study Guide Workbook,* p. 58
____ Vocabulary and Concept Check, *SE,* p. 399
____ Vocabulary Test, *CRM,* p. 446
____ Lesson-by-Lesson Review, *SE,* pp. 399–402
____ Practice Test, *SE,* p. 403
____ Chapter 7 Tests, *CRM,* pp. 433–444
____ Open-Ended Assessment, *CRM,* p. 445
____ Standardized Test Practice, *SE,* pp. 404–405
____ Standardized Test Practice, *CRM,* pp. 451–452
____ Cumulative Review, *CRM,* p. 450
____ *Vocabulary PuzzleMaker CD-ROM*
____ algebra1.com/vocabulary_review
____ algebra1.com/chapter_test
____ algebra1.com/standardized_test
____ *MindJogger Videoquizzes VHS*
____ Unit 2 Test, *CRM,* pp. 453–454

Other Assessment Materials

- *TestCheck and Worksheet Builder CD-ROM*

KEY   SE = Student Edition   TWE = Teacher Wraparound Edition   CRM = Chapter Resource Masters

Lesson Planning Guide (pp. 410–415)

8-1

Teacher's Name _____ Dates _____

Grade _____ Class _____ M Tu W Th F

NCTM Standards
2, 6, 8, 9, 10

Recommended Pacing	
Regular Average	Day 1 of 16
Regular Advanced	Day 1 of 14
Block Average	Day 1 of 8
Block Advanced	Day 1 of 7

Objectives
___ Multiply monomials.
___ Simplify expressions involving powers of monomials.
___ State/local objectives: _____

1 Focus
Materials/Resources Needed _____
___ Building on Prior Knowledge, *TWE*, p. 411
___ 5-Minute Check Transparencies, Lesson 8-1
___ Mathematical Background, *TWE*, p. 408C
___ TeacherWorks CD-ROM

2 Teach
___ In-Class Examples, *TWE*, pp. 411–412
___ *Teaching Algebra with Manipulatives*, p. 134
___ Interactive Chalkboard CD-ROM, Lesson 8-1
___ algebra1.com/extra_examples
___ *Guide to Daily Intervention*, pp. 22–23
___ Daily Intervention, *TWE*, p. 413
___ Study Guide and Intervention, *CRM*, pp. 455–456
___ Reading to Learn Mathematics, *CRM*, p. 459
___ TeacherWorks CD-ROM

3 Practice/Apply
___ Skills Practice, *CRM*, p. 457
___ Practice, *CRM*, p. 458
___ Extra Practice, *SE*, p. 837
___ Differentiated Instruction, *TWE*, p. 412
___ Parent and Student Study Guide Workbook, p. 59
___ Answer Key Transparencies, Lesson 8-1
___ AlgePASS CD-ROM, Lesson 18

Assignment Guide, pp. 413–415, *SE*			
	Objective 1	Objective 2	Other
Basic	15–25 odd	27–37 odd, 43–47 odd, 49–52	55–82
Average	15–25 odd	27–47 odd, 51–54	55–82
Advanced	16–26 even	28–50 even, 54	53, 55–74 (optional: 75–82)
Alternate Assignment			

4 Assess
___ Open-Ended Assessment, *TWE*, p. 415
___ Enrichment, *CRM*, p. 460
___ algebra1.com/self_check_quiz

___ *Closing the Gap for Absent Students*, pp. 16–17

KEY SE = Student Edition   TWE = Teacher Wraparound Edition   CRM = Chapter Resource Masters

© Glencoe/McGraw-Hill   Glencoe Algebra 1

Algebra Activity (p. 416)
A Follow-Up of Lesson 8-1

8

Teacher's Name _____ Dates _____

Grade _____ Class _____ M Tu W Th F

NCTM Standards
2, 3, 6, 7

Recommended Pacing	
Regular Average	Day 2 of 16
Regular Advanced	Day 2 of 14
Block Average	Day 1 of 8
Block Advanced	Day 1 of 7

Objectives
____ Use paper prisms to investigate surface area and volume.
____ State/local objectives: _____

Getting Started
Materials/Resources Needed centimeter grid paper, scissors, tape

Teach
____ *Teaching Algebra with Manipulatives*, p. 135
____ *Glencoe Mathematics Classroom Manipulative Kit*
____ Teaching Strategy, *TWE*, p. 416

Assignment Guide, p. 416, *SE*	
All	1–6
Alternate Assignment	

Assess
____ Study Notebook, *TWE*, p. 416

KEY   *SE* = Student Edition   *TWE* = Teacher Wraparound Edition   *CRM* = Chapter Resource Masters

Lesson Planning Guide (pp. 417–424)

8-2

Teacher's Name _____   Dates _____

Grade _____ Class _____ M Tu W Th F

NCTM Standards
2, 6, 8, 9, 10

Recommended Pacing	
Regular Average	Days 3 & 4 of 16
Regular Advanced	Days 3 & 4 of 14
Block Average	Days 1 & 2 of 8
Block Advanced	Days 1 & 2 of 7

Objectives
____ Simplify expressions involving the quotient of monomials.
____ Simplify expressions containing negative exponents.
____ State/local objectives: _____

1 Focus
Materials/Resources Needed _____

____ Building on Prior Knowledge, *TWE*, p. 418
____ *5-Minute Check Transparencies*, Lesson 8-2
____ Mathematical Background, *TWE*, p. 408C
____ *TeacherWorks CD-ROM*

2 Teach
____ In-Class Examples, *TWE*, pp. 418–420
____ *Interactive Chalkboard CD-ROM*, Lesson 8-2
____ algebra1.com/extra_examples
____ *Guide to Daily Intervention*, pp. 22–23
____ Daily Intervention, *TWE*, p. 421
____ Study Guide and Intervention, *CRM*, pp. 461–462
____ Reading to Learn Mathematics, *CRM*, p. 465
____ *TeacherWorks CD-ROM*
____ Reading Mathematics, *SE*, p. 424

3 Practice/Apply
____ Skills Practice, *CRM*, p. 463
____ Practice, *CRM*, p. 464
____ Extra Practice, *SE*, p. 837
____ Differentiated Instruction, *TWE*, p. 419
____ *Graphing Calculator and Spreadsheet Masters*, p. 61
____ *Parent and Student Study Guide Workbook*, p. 60
____ Answer Key Transparencies, Lesson 8-2
____ *AlgePASS CD-ROM*, Lesson 19

Assignment Guide, pp. 421–423, *SE*			
	Objective 1	Objective 2	Other
Basic	15–21 odd, 39	23–35 odd, 41	40, 42, 47–77
Average	15–21 odd, 39	23–37 odd, 40–44	47–77
Advanced	14–20 even, 38	22–36 even, 44	43, 45–71 (optional: 72–77)
Reading Mathematics	1–3		

Alternate Assignment _____

4 Assess
____ Open-Ended Assessment, *TWE*, p. 423
____ Enrichment, *CRM*, p. 466
____ Assessment, Quiz, *CRM*, p. 423
____ algebra1.com/self_check_quiz

____ Closing the Gap for Absent Students, pp. 16–17

KEY *SE* = Student Edition   *TWE* = Teacher Wraparound Edition   *CRM* = Chapter Resource Masters

© Glencoe/McGraw-Hill   Glencoe Algebra 1

Lesson Planning Guide (pp. 425–430)

8-3

Teacher's Name _____ Dates _____

Grade _____ Class _____ M Tu W Th F

NCTM Standards
1, 6, 8, 9, 10

Recommended Pacing	
Regular Average	Day 5 of 16
Regular Advanced	Day 5 of 14
Block Average	Day 3 of 8
Block Advanced	Day 3 of 7

Objectives

____ Express numbers in scientific notation and standard notation.

____ Find products and quotients of numbers expressed in scientific notation.

____ State/local objectives: _____

1 Focus

Materials/Resources Needed _____
_____

____ Building on Prior Knowledge, *TWE*, p. 427
____ *5-Minute Check Transparencies*, Lesson 8-3
____ Mathematical Background, *TWE*, p. 408C
____ *Prerequisite Skills Masters*, pp. 33–36
____ *TeacherWorks CD-ROM*

2 Teach

____ In-Class Examples, *TWE*, pp. 426–427
____ *Interactive Chalkboard CD-ROM*, Lesson 8-3
____ algebra1.com/extra_examples
____ algebra1.com/data_update
____ algebra1.com/usa_today
____ *Guide to Daily Intervention*, pp. 22–23
____ Study Guide and Intervention, *CRM*, pp. 467–468
____ Reading to Learn Mathematics, *CRM*, p. 471
____ *TeacherWorks CD-ROM*

3 Practice/Apply

____ Skills Practice, *CRM*, p. 469
____ Practice, *CRM*, p. 470
____ Extra Practice, *SE*, p. 837
____ Differentiated Instruction, *TWE*, p. 426
____ School-to-Career Masters, p. 15
____ *Science and Mathematics Lab Manual*, pp. 69–72
____ *Graphing Calculator and Spreadsheet Masters*, p. 62
____ *Parent and Student Study Guide Workbook*, p. 61
____ *Real-World Transparency and Master*
____ *Answer Key Transparencies*, Lesson 8-3
____ *AlgePASS CD-ROM*, Lesson 20
____ algebra1.com/webquest

Assignment Guide, pp. 428–430, *SE*			
	Objective 1	Objective 2	Other
Basic	19–23 odd, 27–39 odd, 43	45–57 odd	60–63, 68–82
Average	19–43 odd	45–59 odd	60–63, 68–82 (optional: 64–67)
Advanced	18–42 even	44–58 even	60–76 (optional: 77–82)
All	Practice Quiz 1 (1–10)		
Alternate Assignment	_____		

4 Assess

____ Practice Quiz 1, *SE*, p. 430
____ Open-Ended Assessment, *TWE*, p. 430
____ Enrichment, *CRM*, p. 472
____ algebra1.com/self_check_quiz

____ *Closing the Gap for Absent Students*, pp. 16–17

KEY	*SE* = Student Edition	*TWE* = Teacher Wraparound Edition	*CRM* = Chapter Resource Masters

© Glencoe/McGraw-Hill   Glencoe Algebra 1

Algebra Activity (p. 431)
A Preview of Lesson 8-4

8

Teacher's Name _____ Dates _____

Grade _____ Class _____ M Tu W Th F

NCTM Standards
2, 10

Recommended Pacing	
Regular Average	Day 6 of 16
Regular Advanced	Day 6 of 14
Block Average	Day 3 of 8
Block Advanced	Day 3 of 7

Objectives
___ Use algebra tiles to model polynomials.
___ State/local objectives: _____

Getting Started
Materials/Resources Needed __algebra tiles__ _____

Teach
___ *Teaching Algebra with Manipulatives*, p. 136
___ *Glencoe Mathematics Classroom Manipulative Kit*
___ Teaching Strategy, *TWE*, p. 431

Assignment Guide, p. 431, *SE*	
All	1–9
Alternate Assignment	

Assess
___ Study Notebook, *TWE*, p. 431

KEY *SE* = Student Edition   *TWE* = Teacher Wraparound Edition   *CRM* = Chapter Resource Masters

© Glencoe/McGraw-Hill   *Glencoe Algebra 1*

Lesson Planning Guide (pp. 432–436)

8-4

Teacher's Name _____ Dates _____

Grade _____ Class _____ M Tu W Th F

NCTM Standards
2, 6, 8, 9, 10

Recommended Pacing	
Regular Average	Day 7 of 16
Regular Advanced	Day 6 of 14
Block Average	Day 4 of 8
Block Advanced	Day 3 of 7

Objectives
___ Find the degree of a polynomial.
___ Arrange the terms of a polynomial in ascending or descending order.
___ State/local objectives: _____

1 Focus
Materials/Resources Needed _____
_____
___ 5-Minute Check Transparencies, Lesson 8-4
___ Mathematical Background, TWE, p. 408D
___ TeacherWorks CD-ROM

2 Teach
___ In-Class Examples, TWE, pp. 433–434
___ Interactive Chalkboard CD-ROM, Lesson 8-4
___ algebra1.com/extra_examples
___ Guide to Daily Intervention, pp. 22–23
___ Daily Intervention, TWE, p. 433
___ Study Guide and Intervention, CRM, pp. 473–474
___ Reading to Learn Mathematics, CRM, p. 477
___ TeacherWorks CD-ROM

3 Practice/Apply
___ Skills Practice, CRM, p. 475
___ Practice, CRM, p. 476
___ Extra Practice, SE, p. 838
___ Differentiated Instruction, TWE, p. 434
___ Graphing Calculator and Spreadsheet Masters, p. 37
___ Parent and Student Study Guide Workbook, p. 62
___ Answer Key Transparencies, Lesson 8-4
___ WebQuest and Projects Resources, p. 45

Assignment Guide, pp. 434–436, SE			
	Objective 1	Objective 2	Other
Basic	15–35 odd, 53	37–51 odd	57–76
Average	15–35 odd, 53, 55	37–51 odd	56–76
Advanced	16–36 even, 54	38–52 even	55–71 (optional: 72–76)

Alternate Assignment _____

4 Assess
___ Open-Ended Assessment, TWE, p. 436
___ Enrichment, CRM, p. 478
___ Assessment, Mid-Chapter Test, CRM, p. 519
___ Assessment, Quiz, CRM, p. 517
___ algebra1.com/self_check_quiz

___ Closing the Gap for Absent Students, pp. 16–17

KEY SE = Student Edition   TWE = Teacher Wraparound Edition   CRM = Chapter Resource Masters

© Glencoe/McGraw-Hill — Glencoe Algebra 1

Algebra Activity (pp. 437–438)
A Preview of Lesson 8-5

8

Teacher's Name _____ Dates _____

Grade _____ Class _____ M Tu W Th F

NCTM Standards
2, 10

Recommended Pacing	
Regular Average	Day 8 of 16
Regular Advanced	Day 7 of 14
Block Average	Day 4 of 8
Block Advanced	Day 4 of 7

Objectives
____ Use algebra tiles to add and subtract polynomials.
____ State/local objectives: _____

Getting Started
Materials/Resources Needed __algebra tiles__ _____

Teach
____ *Teaching Algebra with Manipulatives*, p. 137
____ *Glencoe Mathematics Classroom Manipulative Kit*
____ Teaching Strategy, *TWE*, p. 437

Assignment Guide, p. 438, *SE*	
All	1–7
Alternate Assignment	

Assess
____ Study Notebook, *TWE*, p. 438

KEY   *SE* = Student Edition   *TWE* = Teacher Wraparound Edition   *CRM* = Chapter Resource Masters

© Glencoe/McGraw-Hill   Glencoe Algebra 1

Lesson Planning Guide (pp. 439–443)

8-5

Teacher's Name _____ Dates _____

Grade _____ Class _____ M Tu W Th F

NCTM Standards
2, 6, 8, 9, 10

Recommended Pacing	
Regular Average	Day 9 of 16
Regular Advanced	Day 8 of 14
Block Average	Day 5 of 8
Block Advanced	Day 4 of 7

Objectives
___ Add polynomials.
___ Subtract polynomials.
___ State/local objectives: _____

1 Focus
Materials/Resources Needed _____
___ *5-Minute Check Transparencies,* Lesson 8-5
___ Mathematical Background, *TWE,* p. 408D
___ *TeacherWorks CD-ROM*

2 Teach
___ In-Class Examples, *TWE,* p. 440
___ *Teaching Algebra with Manipulatives,* pp. 138–141
___ *Interactive Chalkboard CD-ROM,* Lesson 8-5
___ algebra1.com/extra_examples
___ algebra1.com/careers
___ *Guide to Daily Intervention,* pp. 22–23
___ Daily Intervention, *TWE,* p. 441
___ Study Guide and Intervention, *CRM,* pp. 479–480
___ Reading to Learn Mathematics, *CRM,* p. 483
___ *TeacherWorks CD-ROM*

3 Practice/Apply
___ Skills Practice, *CRM,* p. 481
___ Practice, *CRM,* p. 482
___ Extra Practice, *SE,* p. 838
___ Differentiated Instruction, *TWE,* p. 441
___ *Parent and Student Study Guide Workbook,* p. 63
___ *Answer Key Transparencies,* Lesson 8-5

Assignment Guide, pp. 441–443, SE			
	Objective 1	Objective 2	Other
Basic	13–17 odd, 27	19–25 odd	32, 33, 41, 44–68
Average	13–17 odd, 27, 34–38	19–25 odd, 29, 31	41–68
Advanced	12–16 even, 26, 36–40	18–24 even, 28, 30	41–62 (optional: 63–68)
Alternate Assignment			

4 Assess
___ Open-Ended Assessment, *TWE,* p. 443
___ Enrichment, *CRM,* p. 484
___ algebra1.com/self_check_quiz

___ *Closing the Gap for Absent Students,* pp. 16–17

KEY *SE* = Student Edition   *TWE* = Teacher Wraparound Edition   *CRM* = Chapter Resource Masters

© Glencoe/McGraw-Hill   Glencoe Algebra 1

Lesson Planning Guide (pp. 444–449)

8-6

Teacher's Name _____ Dates _____
Grade _____ Class _____ M Tu W Th F

NCTM Standards
2, 6, 8, 9, 10

Recommended Pacing	
Regular Average	Days 10 & 11 of 16
Regular Advanced	Day 9 of 14
Block Average	Days 5 & 6 of 8
Block Advanced	Day 5 of 7

Objectives
____ Find the product of a monomial and a polynomial.
____ Solve equations involving polynomials.
____ State/local objectives: _____

1 Focus
Materials/Resources Needed _____

____ *5-Minute Check Transparencies,* Lesson 8-6
____ *Mathematical Background, TWE,* p. 408D
____ *TeacherWorks CD-ROM*

2 Teach
____ In-Class Examples, *TWE,* p. 445
____ *Teaching Algebra with Manipulatives,* pp. 142–143
____ *Interactive Chalkboard CD-ROM,* Lesson 8-6
____ algebra1.com/extra_examples
____ *Guide to Daily Intervention,* pp. 22–23
____ Study Guide and Intervention, *CRM,* pp. 485–486
____ Reading to Learn Mathematics, *CRM,* p. 489
____ *TeacherWorks CD-ROM*

3 Practice/Apply
____ Skills Practice, *CRM,* p. 487
____ Practice, *CRM,* p. 488
____ *Extra Practice, SE,* p. 838
____ *Differentiated Instruction, TWE,* p. 445
____ *School-to-Career Masters,* p. 16
____ *Parent and Student Study Guide Workbook,* p. 64
____ *Answer Key Transparencies,* Lesson 8-6
____ *AlgePASS CD-ROM,* Lesson 21

Assignment Guide, pp. 446–449, *SE*			
	Objective 1	Objective 2	Other
Basic	15–35 odd, 51	39–45 odd, 49, 50	55–57, 63–87
Average	15–37 odd, 51	39–47 odd	53, 55–57, 63–87
Advanced	16–38 even, 58–61	40–48 even, 62	55–57, 63–81 (optional: 82–87)
All	Practice Quiz 2 (1–10)		
Alternate Assignment			

4 Assess
____ Practice Quiz 2, *SE,* p. 449
____ Open-Ended Assessment, *TWE,* p. 449
____ Enrichment, *CRM,* p. 490
____ Assessment, Quiz, *CRM,* p. 518
____ algebra1.com/self_check_quiz

____ *Closing the Gap for Absent Students,* pp. 16–17

KEY *SE* = Student Edition *TWE* = Teacher Wraparound Edition *CRM* = Chapter Resource Masters

© Glencoe/McGraw-Hill   Glencoe Algebra 1

Algebra Activity (pp. 450–451)
A Preview of Lesson 8-7

Teacher's Name _____ Dates _____

Grade _____ Class _____ M Tu W Th F

NCTM Standards
2, 10

Recommended Pacing	
Regular Average	Day 12 of 16
Regular Advanced	Day 10 of 14
Block Average	Day 6 of 8
Block Advanced	Day 5 of 7

Objectives
___ Use algebra tiles to find the product of two binomials.
___ State/local objectives: _____

Getting Started
Materials/Resources Needed __algebra tiles, product mat_____

Teach
___ *Teaching Algebra with Manipulatives*, p. 144
___ *Glencoe Mathematics Classroom Manipulative Kit*
___ Teaching Strategy, *TWE*, p. 450

Assignment Guide, p. 451, SE	
All	1–7
Alternate Assignment	

Assess
___ Study Notebook, *TWE*, p. 451

KEY *SE* = Student Edition   *TWE* = Teacher Wraparound Edition   *CRM* = Chapter Resource Masters

Lesson Planning Guide (pp. 452–457)

8-7

Teacher's Name _____ Dates _____

Grade _____ Class _____ M Tu W Th F

NCTM Standards
2, 6, 8, 9, 10

Recommended Pacing	
Regular Average	Day 13 of 16
Regular Advanced	Day 11 of 14
Block Average	Day 7 of 8
Block Advanced	Day 6 of 7

Objectives
___ Multiply two binomials by using the FOIL method.
___ Multiply two polynomials by using the Distributive Property.
___ State/local objectives: _____

1 Focus
Materials/Resources Needed _____
_____
___ Building on Prior Knowledge, *TWE*, p. 452
___ *5-Minute Check Transparencies*, Lesson 8-7
___ Mathematical Background, *TWE*, p. 408D
___ *TeacherWorks CD-ROM*

2 Teach
___ In-Class Examples, *TWE*, pp. 453–454
___ *Teaching Algebra with Manipulatives*, pp. 145–148
___ *Interactive Chalkboard CD-ROM*, Lesson 8-7
___ algebra1.com/extra_examples
___ *Guide to Daily Intervention*, pp. 22–23
___ Study Guide and Intervention, *CRM*, pp. 491–492
___ Reading to Learn Mathematics, *CRM*, p. 495
___ *TeacherWorks CD-ROM*
___ *Multimedia Applications Masters*

3 Practice/Apply
___ Skills Practice, *CRM*, p. 493
___ Practice, *CRM*, p. 494
___ Extra Practice, *SE*, p. 839
___ Differentiated Instruction, *TWE*, p. 453
___ *Graphing Calculator and Spreadsheet Masters*, p. 38
___ *Parent and Student Study Guide Workbook*, p. 65
___ Answer Key Transparencies, Lesson 8-7
___ *AlgePASS CD-ROM*, Lesson 22
___ *WebQuest and Projects Resources*, p. 46

Assignment Guide, pp. 455–457, *SE*			
	Objective 1	Objective 2	Other
Basic	13–29 odd, 39, 41	31–37 odd, 45–47	54–77
Average	13–29 odd, 39, 41, 49–51	31–37 odd, 43, 45–47	53–77
Advanced	14–30 even, 40, 42, 49–52	32–38 even, 44	54–71 (optional: 72–77)
Alternate Assignment			

4 Assess
___ Open-Ended Assessment, *TWE*, p. 457
___ Enrichment, *CRM*, p. 496
___ algebra1.com/self_check_quiz

___ Closing the Gap for Absent Students, pp. 16–17

KEY *SE* = Student Edition *TWE* = Teacher Wraparound Edition *CRM* = Chapter Resource Masters

© Glencoe/McGraw-Hill Glencoe Algebra 1

Lesson Planning Guide (pp. 458–463)

8-8

Teacher's Name _____ Dates _____

Grade _____ Class _____ M Tu W Th F

NCTM Standards
2, 6, 8, 9, 10

Recommended Pacing	
Regular Average	Day 14 of 16
Regular Advanced	Day 12 of 14
Block Average	Day 7 of 8
Block Advanced	Day 6 of 7

Objectives

____ Find squares of sums and differences.
____ Find the product of a sum and a difference.
____ State/local objectives: _____

---

1 Focus

Materials/Resources Needed _____

____ *5-Minute Check Transparencies*, Lesson 8-8
____ Mathematical Background, *TWE*, p. 408D
____ *TeacherWorks CD-ROM*

2 Teach

____ In-Class Examples, *TWE*, pp. 459–460
____ *Teaching Algebra with Manipulatives*, pp. 149–151
____ *Interactive Chalkboard CD-ROM*, Lesson 8-8
____ algebra1.com/extra_examples
____ algebra1.com/careers
____ *Guide to Daily Intervention*, pp. 22–23
____ Study Guide and Intervention, *CRM*, pp. 497–498
____ *Reading to Learn Mathematics*, *CRM*, p. 501
____ *TeacherWorks CD-ROM*

3 Practice/Apply

____ Skills Practice, *CRM*, p. 499
____ Practice, *CRM*, p. 500
____ Extra Practice, *SE*, p. 839
____ Differentiated Instruction, *TWE*, p. 460
____ *Parent and Student Study Guide Workbook*, p. 66
____ *Answer Key Transparencies*, Lesson 8-8
____ *AlgePASS CD-ROM*, Lesson 23

Assignment Guide, pp. 461–463, *SE*			
	Objective 1	Objective 2	Other
Basic	13, 15, 19, 21, 25, 27, 31, 35, 39–44	17, 23, 29, 33	48–50, 52–70
Average	13, 15, 19, 21, 25, 27, 31, 35, 39–44	17, 23, 29, 33, 37, 47	48–50, 52–70 (optional: 51)
Advanced	14, 16, 20, 22, 26, 28, 32, 36, 45, 46	18, 24, 30, 34, 38, 47	48–70
Alternate Assignment			

4 Assess

____ Open-Ended Assessment, *TWE*, p. 463
____ Enrichment, *CRM*, p. 502
____ Assessment, Quiz, *CRM*, p. 518
____ algebra1.com/self_check_quiz

____ *Closing the Gap for Absent Students*, pp. 16–17

KEY   *SE* = Student Edition   *TWE* = Teacher Wraparound Edition   *CRM* = Chapter Resource Masters

© Glencoe/McGraw-Hill   Glencoe Algebra 1

8 Review and Testing (pp. 464–471)

Teacher's Name _____ Dates _____

Grade _____ Class _____ M Tu W Th F

Recommended Pacing	
Regular Average	Days 15 & 16 of 16
Regular Advanced	Days 13 & 14 of 14
Block Average	Day 8 of 8
Block Advanced	Day 7 of 7

Assess

____ *Parent and Student Study Guide Workbook,* p. 67
____ Vocabulary and Concept Check, *SE,* p. 464
____ Vocabulary Test, *CRM,* p. 516
____ Lesson-by-Lesson Review, *SE,* pp. 464–468
____ Practice Test, *SE,* p. 469
____ Chapter 8 Tests, *CRM,* pp. 503–514
____ Open-Ended Assessment, *CRM,* p. 515
____ Standardized Test Practice, *SE,* pp. 470–471
____ Standardized Test Practice, *CRM,* pp. 521–522
____ Cumulative Review, *CRM,* p. 520
____ *Vocabulary PuzzleMaker CD-ROM*
____ algebra1.com/vocabulary_review
____ algebra1.com/chapter_test
____ algebra1.com/standardized_test
____ *MindJogger Videoquizzes VHS*

Other Assessment Materials

- *TestCheck and Worksheet Builder CD-ROM*

KEY   *SE* = Student Edition   *TWE* = Teacher Wraparound Edition   *CRM* = Chapter Resource Masters

Lesson Planning Guide (pp. 474–479)

9-1

Teacher's Name _____ Dates _____

Grade _____ Class _____ M Tu W Th F

NCTM Standards
1, 2, 6, 8, 9, 10

Recommended Pacing	
Regular Average	Day 1 of 14
Regular Advanced	Day 1 of 12
Block Average	Day 1 of 8
Block Advanced	Day 1 of 6

Objectives
____ Find prime factorizations of integers and monomials.
____ Find the greatest common factors of integers and monomials.
____ State/local objectives: _____

1 Focus
Materials/Resources Needed _____
_____
____ 5-Minute Check Transparencies, Lesson 9-1
____ Mathematical Background, TWE, p. 472C
____ Prerequisite Skills Masters, pp. 13–14
____ TeacherWorks CD-ROM

2 Teach
____ In-Class Examples, TWE, pp. 475–476
____ Interactive Chalkboard CD-ROM, Lesson 9-1
____ algebra1.com/extra_examples
____ Guide to Daily Intervention, pp. 24–25
____ Study Guide and Intervention, CRM, pp. 523–524
____ Reading to Learn Mathematics, CRM, p. 527
____ TeacherWorks CD-ROM

3 Practice/Apply
____ Skills Practice, CRM, p. 525
____ Practice, CRM, p. 526
____ Extra Practice, SE, p. 839
____ Differentiated Instruction, TWE, p. 475
____ Parent and Student Study Guide Workbook, p. 68
____ Answer Key Transparencies, Lesson 9-1
____ algebra1.com/webquest

Assignment Guide, pp. 477–479, SE			
	Objective 1	Objective 2	Other
Basic	21–27 odd, 33–39 odd	49–61 odd	29–31, 41–47 odd, 68–86
Average	21–27 odd, 33–39 odd	49–61 odd	29, 41–47 odd, 63–66, 68–86
Advanced	20–26 even, 32–38 even	48–60 even	28, 40–46 even, 62–80 (optional: 81–86)

Alternate Assignment _____

4 Assess
____ Open-Ended Assessment, TWE, p. 479
____ Enrichment, CRM, p. 528
____ algebra1.com/self_check_quiz

____ Closing the Gap for Absent Students, pp. 18–19

KEY SE = Student Edition   TWE = Teacher Wraparound Edition   CRM = Chapter Resource Masters

© Glencoe/McGraw-Hill   Glencoe Algebra 1

Algebra Activity (p. 480)
A Preview of Lesson 9-2

9

Teacher's Name _____ Dates _____

Grade _____ Class _____ M Tu W Th F

NCTM Standards
2, 6, 8, 10

Recommended Pacing	
Regular Average	Day 2 of 14
Regular Advanced	Day 2 of 12
Block Average	Day 1 of 8
Block Advanced	Day 1 of 6

Objectives
___ Use algebra tiles and a product mat to factor binomials.
___ State/local objectives: _____

Getting Started
Materials/Resources Needed __algebra tiles, product mat__

Teach
___ *Teaching Algebra with Manipulatives*, p. 156
___ *Glencoe Mathematics Classroom Manipulative Kit*
___ Teaching Strategy, *TWE*, p. 480

Assignment Guide, p. 480, *SE*	
All	1–9
Alternate Assignment	

Assess
___ Study Notebook, *TWE*, p. 480

KEY *SE* = Student Edition   *TWE* = Teacher Wraparound Edition   *CRM* = Chapter Resource Masters

© Glencoe/McGraw-Hill   Glencoe Algebra 1

9-2 Lesson Planning Guide (pp. 481–486)

Teacher's Name _____ Dates _____

Grade _____ Class _____ M Tu W Th F

NCTM Standards
2, 6, 8, 9, 10

Recommended Pacing	
Regular Average	Day 3 of 14
Regular Advanced	Day 2 of 12
Block Average	Day 2 of 8
Block Advanced	Day 1 of 6

Objectives
___ Factor polynomials by using the Distributive Property.
___ Solve quadratic equations of the form $ax^2 + bx = 0$.
___ State/local objectives: _____

---

1 Focus
Materials/Resources Needed _____

___ Building on Prior Knowledge, *TWE*, p. 481
___ *5-Minute Check Transparencies*, Lesson 9-2
___ *Mathematical Background*, *TWE*, p. 472C
___ *Prerequisite Skills Masters*, pp. 13–14
___ *TeacherWorks CD-ROM*

2 Teach
___ In-Class Examples, *TWE*, pp. 482–483
___ *Teaching Algebra with Manipulatives*, pp. 157–158
___ *Interactive Chalkboard CD-ROM*, Lesson 9-2
___ algebra1.com/extra_examples
___ algebra1.com/careers
___ *Guide to Daily Intervention*, pp. 24–25
___ Study Guide and Intervention, *CRM*, pp. 529–530
___ *Reading to Learn Mathematics*, *CRM*, p. 533
___ *TeacherWorks CD-ROM*

3 Practice/Apply
___ Skills Practice, *CRM*, p. 531
___ Practice, *CRM*, p. 532
___ Extra Practice, *SE*, p. 840
___ Differentiated Instruction, *TWE*, p. 483
___ *School-to-Career Masters*, p. 17
___ *Parent and Student Study Guide Workbook*, p. 69
___ *Answer Key Transparencies*, Lesson 9-2

Assignment Guide, pp. 484–486, *SE*			
	Objective 1	Objective 2	Other
Basic	17–39 odd	49–59 odd	40–43, 47, 48, 62–81
Average	17–39 odd	49–59 odd	42, 43–47 odd, 61–81
Advanced	16–38 even	48–58 even	44–46, 60, 62–75 (optional: 76–81)
Alternate Assignment			

4 Assess
___ Practice Quiz 1, *SE*, p. 486
___ Open-Ended Assessment, *TWE*, p. 486
___ Enrichment, *CRM*, p. 534

___ Assessment, Quiz, *CRM*, p. 573
___ algebra1.com/self_check_quiz

___ Closing the Gap for Absent Students, pp. 18–19

KEY *SE* = Student Edition   *TWE* = Teacher Wraparound Edition   *CRM* = Chapter Resource Masters

© Glencoe/McGraw-Hill   Glencoe Algebra 1

Algebra Activity (pp. 487–488)
A Preview of Lesson 9-3

9

Teacher's Name _____ Dates _____

Grade _____ Class _____ M Tu W Th F

NCTM Standards
2, 6, 8, 10

Recommended Pacing	
Regular Average	Day 4 of 14
Regular Advanced	Day 3 of 12
Block Average	Day 2 of 8
Block Advanced	Day 2 of 6

Objectives
___ Use algebra tiles to factor trinomials.
___ State/local objectives: _____

Getting Started
Materials/Resources Needed _algebra tiles, product mat_

Teach
___ *Teaching Algebra with Manipulatives*, p. 159
___ *Glencoe Mathematics Classroom Manipulative Kit*
___ Teaching Strategy, *TWE*, p. 487

Assignment Guide, p. 488, *SE*	
All	1–8
Alternate Assignment	

Assess
___ Study Notebook, *TWE*, p. 488

KEY *SE* = Student Edition *TWE* = Teacher Wraparound Edition *CRM* = Chapter Resource Masters

© Glencoe/McGraw-Hill   Glencoe Algebra 1

Lesson Planning Guide (pp. 489–494)

9-3

Teacher's Name _____ Dates _____
Grade _____ Class _____ M Tu W Th F

NCTM Standards
2, 6, 8, 9, 10

Recommended Pacing	
Regular Average	Days 5 & 6 of 14
Regular Advanced	Day 4 of 12
Block Average	Days 3 & 4 of 8
Block Advanced	Day 2 of 6

Objectives
____ Factor trinomials of the form $x^2 + bx + c$.
____ Solve equations of the form $x^2 + bx + c = 0$.
____ State/local objectives: _____

1 Focus
Materials/Resources Needed _____

____ *5-Minute Check Transparencies*, Lesson 9-3
____ Mathematical Background, *TWE*, p. 472D
____ *TeacherWorks CD-ROM*

2 Teach
____ In-Class Examples, *TWE*, pp. 490–491
____ *Teaching Algebra with Manipulatives*, pp. 160–161
____ *Interactive Chalkboard CD-ROM*, Lesson 9-3
____ algebra1.com/extra_examples
____ *Guide to Daily Intervention*, pp. 24–25
____ Daily Intervention, *TWE*, p. 492
____ Study Guide and Intervention, *CRM*, pp. 535–536
____ Reading to Learn Mathematics, *CRM*, p. 539
____ *TeacherWorks CD-ROM*

3 Practice/Apply
____ Skills Practice, *CRM*, p. 537
____ Practice, *CRM*, p. 538
____ Extra Practice, *SE*, p. 840
____ Differentiated Instruction, *TWE*, p. 490
____ *Parent and Student Study Guide Workbook*, p. 70
____ *Answer Key Transparencies*, Lesson 9-3
____ *AlgePASS CD-ROM*, Lessons 24 & 25

Assignment Guide, pp. 492–494, *SE*			
	Objective 1	Objective 2	Other
Basic	17–33 odd	37–51 odd	55, 57–60, 63–65, 70–83
Average	17–33 odd	37–53 odd	35, 55, 57–60, 63–65, 70–83 (optional: 66–69)
Advanced	18–34 even	38–52 even	36, 54, 56–77 (optional: 78–83)
Alternate Assignment			

4 Assess
____ Open-Ended Assessment, *TWE*, p. 494
____ Enrichment, *CRM*, p. 540
____ Assessment, Mid-Chapter Test, *CRM*, p. 575
____ Assessment, Quiz, *CRM*, p. 573
____ algebra1.com/self_check_quiz

____ *Closing the Gap for Absent Students*, pp. 18–19

KEY *SE* = Student Edition *TWE* = Teacher Wraparound Edition *CRM* = Chapter Resource Masters

Lesson Planning Guide (pp. 495–500)

9-4

Teacher's Name _____ Dates _____

Grade _____ Class _____ M Tu W Th F

NCTM Standards
2, 6, 8, 9, 10

Recommended Pacing	
Regular Average	Days 7 & 8 of 14
Regular Advanced	Days 5 & 6 of 12
Block Average	Days 4 & 5 of 8
Block Advanced	Day 3 of 6

Objectives
___ Factor trinomials of the form $ax^2 + bx + c$.
___ Solve equations of the form $ax^2 + bx + c = 0$.
___ State/local objectives: _____

1 Focus
Materials/Resources Needed _____

___ *5-Minute Check Transparencies*, Lesson 9-4
___ Mathematical Background, *TWE*, p. 472D
___ *Prerequisite Skills Masters*, pp. 13–14
___ *TeacherWorks CD-ROM*

2 Teach
___ In-Class Examples, *TWE*, pp. 496–497
___ *Teaching Algebra with Manipulatives*, pp. 162–164
___ *Interactive Chalkboard CD-ROM*, Lesson 9-4
___ algebra1.com/extra_examples
___ *Guide to Daily Intervention*, pp. 24–25
___ Daily Intervention, *TWE*, pp. 497, 498
___ Study Guide and Intervention, *CRM*, pp. 541–542
___ Reading to Learn Mathematics, *CRM*, p. 545
___ *TeacherWorks CD-ROM*

3 Practice/Apply
___ Skills Practice, *CRM*, p. 543
___ Practice, *CRM*, p. 544
___ Extra Practice, *SE*, p. 840
___ Differentiated Instruction, *TWE*, p. 496
___ *Graphing Calculator and Spreadsheet Masters*, p. 39
___ *Parent and Student Study Guide Workbook*, p. 71
___ Answer Key Transparencies, Lesson 9-4
___ *AlgePASS CD-ROM*, Lesson 26

Assignment Guide, pp. 498–500, *SE*			
	Objective 1	Objective 2	Other
Basic	15–29 odd, 33	35–43 odd	49, 50, 53–70
Average	15–33 odd	35–47 odd	49–51, 53–70
Advanced	14–34 even	36–48 even	52–62 (optional: 63–70)
All	Practice Quiz 2 (1–10)		
Alternate Assignment			

4 Assess
___ Practice Quiz 2, *SE*, p. 500
___ Open-Ended Assessment, *TWE*, p. 500
___ Enrichment, *CRM*, p. 546
___ algebra1.com/self_check_quiz

___ *Closing the Gap for Absent Students*, pp. 18–19

KEY *SE* = Student Edition   *TWE* = Teacher Wraparound Edition   *CRM* = Chapter Resource Masters

© Glencoe/McGraw-Hill   Glencoe Algebra 1

Lesson Planning Guide (pp. 501–507)

9-5

Teacher's Name _____ Dates _____

Grade _____ Class _____ M Tu W Th F

NCTM Standards
2, 3, 6, 8, 9, 10

Recommended Pacing	
Regular Average	Days 9 & 10 of 14
Regular Advanced	Days 7 & 8 of 12
Block Average	Days 5 & 6 of 8
Block Advanced	Day 4 of 6

Objectives

___ Factor binomials that are the differences of squares.
___ Solve equations involving the differences of squares.
___ State/local objectives: _____

1 Focus

Materials/Resources Needed _____

___ Building on Prior Knowledge, *TWE*, p. 501
___ *5-Minute Check Transparencies*, Lesson 9-5
___ Mathematical Background, *TWE*, p. 472D
___ *Prerequisite Skills Masters*, pp. 13–14
___ *TeacherWorks CD-ROM*

2 Teach

___ In-Class Examples, *TWE*, pp. 502–503
___ *Teaching Algebra with Manipulatives*, pp. 165–167
___ *Interactive Chalkboard CD-ROM*, Lesson 9-5
___ algebra1.com/extra_examples
___ *Guide to Daily Intervention*, pp. 24–25
___ Daily Intervention, *TWE*, p. 504
___ Study Guide and Intervention, *CRM*, pp. 547–548
___ Reading to Learn Mathematics, *CRM*, p. 551
___ *TeacherWorks CD-ROM*
___ Reading Mathematics, *SE*, p. 507

3 Practice/Apply

___ Skills Practice, *CRM*, p. 549
___ Practice, *CRM*, p. 550
___ Extra Practice, *SE*, p. 841
___ Differentiated Instruction, *TWE*, p. 504
___ *School-to-Career Masters*, p. 18
___ *Science and Mathematics Lab Manual*, pp. 73–78
___ *Graphing Calculator and Spreadsheet Masters*, p. 40
___ *Parent and Student Study Guide Workbook*, p. 72
___ *Answer Key Transparencies*, Lesson 9-5
___ *WebQuest and Projects Resources*, p. 53

Assignment Guide, pp. 504–506, SE			
	Objective 1	Objective 2	Other
Basic	17–31 odd	35–43 odd	46, 47, 49, 51–70
Average	17–33 odd	35–43 odd	46, 47, 49, 51–70
Advanced	16–32 even	34–44 even	46–50 even, 51–64 (optional: 65–70)
Reading Mathematics	1–4		
Alternate Assignment			

4 Assess

___ Open-Ended Assessment, *TWE*, p. 506
___ Enrichment, *CRM*, p. 552
___ Assessment, Quiz, *CRM*, p. 574
___ algebra1.com/self_check_quiz

___ Closing the Gap for Absent Students, pp. 18–19

KEY *SE* = Student Edition   *TWE* = Teacher Wraparound Edition   *CRM* = Chapter Resource Masters

© Glencoe/McGraw-Hill   Glencoe Algebra 1

Lesson Planning Guide (pp. 508–514)

9-6

Teacher's Name _____ Dates _____

Grade _____ Class _____ M Tu W Th F

NCTM Standards
2, 6, 8, 9, 10

Recommended Pacing	
Regular Average	Days 11 & 12 of 14
Regular Advanced	Days 9 & 10 of 12
Block Average	Days 6 & 7 of 8
Block Advanced	Day 5 of 6

Objectives
___ Factor perfect square trinomials.
___ Solve equations involving perfect squares.
___ State/local objectives: _____

1 Focus
Materials/Resources Needed _____
_____
___ Building on Prior Knowledge, *TWE*, p. 509
___ 5-Minute Check Transparencies, Lesson 9-6
___ Mathematical Background, *TWE*, p. 472D
___ *TeacherWorks CD-ROM*

2 Teach
___ In-Class Examples, *TWE*, pp. 509–511
___ *Teaching Algebra with Manipulatives*, pp. 168–172
___ *Interactive Chalkboard CD-ROM*, Lesson 9-6
___ algebra1.com/extra_examples
___ *Guide to Daily Intervention*, pp. 24–25
___ Study Guide and Intervention, *CRM*, pp. 553–554
___ Reading to Learn Mathematics, *CRM*, p. 557
___ *TeacherWorks CD-ROM*
___ *Multimedia Applications Masters*

3 Practice/Apply
___ Skills Practice, *CRM*, p. 555
___ Practice, *CRM*, p. 556
___ Extra Practice, *SE*, p. 841
___ Differentiated Instruction, *TWE*, p. 510
___ *Parent and Student Study Guide Workbook*, p. 73
___ *Real-World Transparency and Master*
___ Answer Key Transparencies, Lesson 9-6
___ *AlgePASS CD-ROM*, Lesson 27
___ *WebQuest and Projects Resources*, p. 53

Assignment Guide, pp. 512–514, *SE*			
	Objective 1	Objective 2	Other
Basic	17–21 odd	43–47 odd	25–39 odd, 49–53 odd, 55, 56, 60–80
Average	17–21 odd	43–47 odd	23–41 odd, 49–53 odd, 55–59, 60–80
Advanced	18–22 even	44–48 even	24–42 even, 50–54 even, 57, 58, 60–80

Alternate Assignment _____

4 Assess
___ Open-Ended Assessment, *TWE*, p. 514
___ Enrichment, *CRM*, p. 558
___ Assessment, Quiz, *CRM*, p. 574
___ algebra1.com/self_check_quiz

___ *Closing the Gap for Absent Students*, pp. 18–19

KEY *SE* = Student Edition   *TWE* = Teacher Wraparound Edition   *CRM* = Chapter Resource Masters

© Glencoe/McGraw-Hill   Glencoe Algebra 1

Review and Testing (pp. 515–521)

9

Teacher's Name _____ Dates _____

Grade _____ Class _____ M Tu W Th F

Recommended Pacing	
Regular Average	Days 13 & 14 of 14
Regular Advanced	Days 11 & 12 of 12
Block Average	Days 7 & 8 of 8
Block Advanced	Day 6 of 6

Assess

____ *Parent and Student Study Guide Workbook*, p. 74
____ Vocabulary and Concept Check, *SE*, p. 515
____ Vocabulary Test, *CRM*, p. 572
____ Lesson-by-Lesson Review, *SE*, pp. 515–518
____ Practice Test, *SE*, p. 519
____ Chapter 9 Tests, *CRM*, pp. 559–570
____ Open-Ended Assessment, *CRM*, p. 571
____ Standardized Test Practice, *SE*, pp. 520–521
____ Standardized Test Practice, *CRM*, pp. 577–578
____ Cumulative Review, *CRM*, p. 576
____ *Vocabulary PuzzleMaker CD-ROM*
____ algebra1.com/vocabulary_review
____ algebra1.com/chapter_test
____ algebra1.com/standardized_test
____ *MindJogger Videoquizzes VHS*

Other Assessment Materials

- *TestCheck and Worksheet Builder CD-ROM*

KEY   *SE* = Student Edition   *TWE* = Teacher Wraparound Edition   *CRM* = Chapter Resource Masters

10-1 Lesson Planning Guide (pp. 524–530)

Teacher's Name _____ Dates _____

Grade _____ Class _____ M Tu W Th F

NCTM Standards
2, 6, 8, 9, 10

Recommended Pacing	
Regular Average	Day 1 of 16
Regular Advanced	Day 1 of 16
Block Average	Day 1 of 9
Block Advanced	Day 1 of 8

Objectives
___ Graph quadratic functions.
___ Find the equation of the axis of symmetry and the coordinates of the vertex of a parabola.
___ State/local objectives: _____

1 Focus
Materials/Resources Needed _____
_____

___ Building on Prior Knowledge, *TWE*, p. 524
___ 5-Minute Check Transparencies, Lesson 10-1
___ Mathematical Background, *TWE*, p. 522C
___ TeacherWorks CD-ROM

2 Teach
___ In-Class Examples, *TWE*, pp. 525–527
___ *Teaching Algebra with Manipulatives*, pp. 176–177
___ *Interactive Chalkboard CD-ROM*, Lesson 10-1
___ algebra1.com/extra_examples
___ *Guide to Daily Intervention*, pp. 26–27
___ Study Guide and Intervention, *CRM*, pp. 579–580
___ Reading to Learn Mathematics, *CRM*, p. 583
___ *TeacherWorks CD-ROM*

3 Practice/Apply
___ Skills Practice, *CRM*, p. 581
___ Practice, *CRM*, p. 582
___ Extra Practice, *SE*, p. 841
___ Differentiated Instruction, *TWE*, p. 526
___ *Graphing Calculator and Spreadsheet Masters*, p. 63
___ *Parent and Student Study Guide Workbook*, p. 75
___ Answer Key Transparencies, Lesson 10-1
___ AlgePASS CD-ROM, Lesson 28
___ WebQuest and Projects Resources, p. 54

Assignment Guide, pp. 528–530, *SE*			
	Objective 1	Objective 2	Other
Basic	11–15 odd	17–29 odd, 37, 39, 51	38, 40, 50, 52, 53, 60–80
Average	11–15 odd	17–37 odd, 39–43, 50, 51	52, 53, 60–80 (optional: 54–59)
Advanced	10–14 even	16–38 even, 44–51	52–74 (optional: 75–80)
Alternate Assignment			

4 Assess
___ Open-Ended Assessment, *TWE*, p. 530
___ Enrichment, *CRM*, p. 584
___ algebra1.com/self_check_quiz

___ Closing the Gap for Absent Students, pp. 20–21

KEY *SE* = Student Edition   *TWE* = Teacher Wraparound Edition   *CRM* = Chapter Resource Masters

Graphing Calculator Investigation (pp. 531-532)
A Follow-Up of Lesson 10-1

Teacher's Name _____ Dates _____

Grade _____ Class _____ M Tu W Th F

NCTM Standards
2, 6, 7, 8

Recommended Pacing	
Regular Average	Day 2 of 16
Regular Advanced	Day 2 of 16
Block Average	Day 2 of 9
Block Advanced	Day 2 of 8

Objectives
___ Use a graphing calculator to explore families of quadratic graphs.
___ State/local objectives: _____

Getting Started
Materials/Resources Needed _____

Teach
___ *Graphing Calculator and Spreadsheet Masters*, p. 64
___ algebra1.com/other_calculator_keystrokes

Assignment Guide, p. 532, *SE*	
All	1–12
Alternate Assignment	

Assess

KEY *SE* = Student Edition   *TWE* = Teacher Wraparound Edition   *CRM* = Chapter Resource Masters

© Glencoe/McGraw-Hill   Glencoe Algebra 1

Lesson Planning Guide (pp. 533–538)

10-2

Teacher's Name _____ Dates _____

Grade _____ Class _____ M Tu W Th F

NCTM Standards
2, 6, 8, 9, 10

Recommended Pacing	
Regular Average	Days 3 & 4 of 16
Regular Advanced	Days 3 & 4 of 16
Block Average	Days 2 & 3 of 9
Block Advanced	Day 2 of 8

Objectives
___ Solve quadratic equations by graphing.
___ Estimate solutions of quadratic equations by graphing.
___ State/local objectives: _____

1 Focus
Materials/Resources Needed _____
_____
___ Building on Prior Knowledge, *TWE*, p. 533
___ *5-Minute Check Transparencies*, Lesson 10-2
___ Mathematical Background, *TWE*, p. 522C
___ *TeacherWorks CD-ROM*

2 Teach
___ In-Class Examples, *TWE*, pp. 534–535
___ *Interactive Chalkboard CD-ROM*, Lesson 10-2
___ algebra1.com/extra_examples
___ *Guide to Daily Intervention*, pp. 26–27
___ Daily Intervention, *TWE*, p. 534
___ Study Guide and Intervention, *CRM*, pp. 585–586
___ Reading to Learn Mathematics, *CRM*, p. 589
___ *TeacherWorks CD-ROM*
___ *Multimedia Applications Masters*

3 Practice/Apply
___ Skills Practice, *CRM*, p. 587
___ Practice, *CRM*, p. 588
___ Extra Practice, *SE*, p. 842
___ Differentiated Instruction, *TWE*, p. 535
___ *School-to-Career Masters*, p. 19
___ *Graphing Calculator and Spreadsheet Masters*, p. 65
___ *Parent and Student Study Guide Workbook*, p. 76
___ *Answer Key Transparencies*, Lesson 10-2
___ *WebQuest and Projects Resources*, p. 54
___ algebra1.com/webquest

Assignment Guide, pp. 536–538, *SE*			
	Objective 1	Objective 2	Other
Basic	11–19 odd, 35–37, 39	21–33 odd	47–50, 53–68
Average	11–19 odd, 35–39	21–33 odd, 41, 42	47–50, 53–68 (optional: 51, 52)
Advanced	12–20 even, 35, 36, 38, 40	22–34 even, 41–46	47–62 (optional: 63–68)
Alternate Assignment			

4 Assess
___ Open-Ended Assessment, *TWE*, p. 538
___ Enrichment, *CRM*, p. 590
___ Assessment, Quiz, *CRM*, p. 635
___ algebra1.com/self_check_quiz

___ *Closing the Gap for Absent Students*, pp. 20–21

KEY *SE* = Student Edition   *TWE* = Teacher Wraparound Edition   *CRM* = Chapter Resource Masters

© Glencoe/McGraw-Hill   Glencoe Algebra 1

Lesson Planning Guide (pp. 539–544)

10-3

Teacher's Name _____ Dates _____

Grade _____ Class _____ M Tu W Th F

NCTM Standards
2, 6, 8, 9, 10

Recommended Pacing	
Regular Average	Day 5 of 16
Regular Advanced	Day 5 of 16
Block Average	Day 3 of 9
Block Advanced	Day 3 of 8

Objectives
___ Solve quadratic equations by finding the square root.
___ Solve quadratic equations by completing the square.
___ State/local objectives: _____

1 Focus
Materials/Resources Needed _____
___ Building on Prior Knowledge, *TWE*, p. 539
___ 5-Minute Check Transparencies, Lesson 10-3
___ Mathematical Background, *TWE*, p. 522C
___ TeacherWorks CD-ROM

2 Teach
___ In-Class Examples, *TWE*, pp. 540–541
___ *Teaching Algebra with Manipulatives*, pp. 178–179
___ *Interactive Chalkboard CD-ROM*, Lesson 10-3
___ algebra1.com/extra_examples
___ algebra1.com/careers
___ algebra1.com/data_update
___ *Guide to Daily Intervention*, pp. 26–27
___ Daily Intervention, *TWE*, p. 540
___ Study Guide and Intervention, *CRM*, pp. 591–592
___ Reading to Learn Mathematics, *CRM*, p. 595
___ TeacherWorks CD-ROM

3 Practice/Apply
___ Skills Practice, *CRM*, p. 593
___ Practice, *CRM*, p. 594
___ Extra Practice, *SE*, p. 842
___ Differentiated Instruction, *TWE*, p. 541
___ *Parent and Student Study Guide Workbook*, p. 77
___ Answer Key Transparencies, Lesson 10-3
___ AlgePASS CD-ROM, Lesson 29

Assignment Guide, pp. 542–544, *SE*			
	Objective 1	Objective 2	Other
Basic	15–19 odd	21–25 odd, 29–45 odd, 49, 51	53–75
Average	15–19 odd	21–51 odd	53–75
Advanced	16–20 even	22–52 even	50–51, 53–71 (optional: 72–75)
All	Practice Quiz 1 (1–10)		
Alternate Assignment			

4 Assess
___ Practice Quiz 1, *SE*, p. 544
___ Open-Ended Assessment, *TWE*, p. 544
___ Enrichment, *CRM*, p. 596
___ algebra1.com/self_check_quiz

___ *Closing the Gap for Absent Students*, pp. 20–21

KEY *SE* = Student Edition   *TWE* = Teacher Wraparound Edition   *CRM* = Chapter Resource Masters

© Glencoe/McGraw-Hill   Glencoe Algebra 1

Graphing Calculator Investigation (p. 545)
A Follow-Up of Lesson 10-3

10

Teacher's Name _____ Dates _____

Grade _____ Class _____ M Tu W Th F

NCTM Standards
2, 6, 7, 8

Recommended Pacing	
Regular Average	Day 6 of 16
Regular Advanced	Day 6 of 16
Block Average	Day 4 of 9
Block Advanced	Day 3 of 8

Objectives
____ Using a graphing calculator to graph quadratic functions in vertex form.
____ State/local objectives: _____

Getting Started
Materials/Resources Needed _____
_____
_____

Teach
____ *Graphing Calculator and Spreadsheet Masters*, p. 66
____ algebra1.com/other_calculator_keystrokes

Assignment Guide, p. 545, SE	
All	1–5
Alternate Assignment	

Assess

KEY *SE* = Student Edition   *TWE* = Teacher Wraparound Edition   *CRM* = Chapter Resource Masters

Lesson Planning Guide (pp. 546–552)

10-4

Teacher's Name _____ Dates _____

Grade _____ Class _____ M Tu W Th F

NCTM Standards
2, 6, 8, 9, 10

Recommended Pacing	
Regular Average	Day 7 of 16
Regular Advanced	Day 7 of 16
Block Average	Days 4 & 5 of 9
Block Advanced	Day 4 of 8

Objectives

____ Solve quadratic equations by using the Quadratic Formula.

____ Use the discriminant to determine the number of solutions for a quadratic equation.

____ State/local objectives: _____

1 Focus

Materials/Resources Needed _____

____ 5-Minute Check Transparencies, Lesson 10-4
____ Mathematical Background, *TWE*, p. 522D
____ *TeacherWorks* CD-ROM

2 Teach

____ In-Class Examples, *TWE*, pp. 547–549
____ *Interactive Chalkboard* CD-ROM, Lesson 10-4
____ algebra1.com/extra_examples
____ Guide to Daily Intervention, pp. 26–27
____ Daily Intervention, *TWE*, p. 550
____ Study Guide and Intervention, *CRM*, pp. 597–598
____ Reading to Learn Mathematics, *CRM*, p. 601
____ *TeacherWorks* CD-ROM

3 Practice/Apply

____ Skills Practice, *CRM*, p. 599
____ Practice, *CRM*, p. 600
____ Extra Practice, *SE*, p. 842
____ Differentiated Instruction, *TWE*, p. 547
____ School-to-Career Masters, p. 20
____ Parent and Student Study Guide Workbook, p. 78
____ Answer Key Transparencies, Lesson 10-4
____ *AlgePASS* CD-ROM, Lessons 30 & 31

Assignment Guide, pp. 550–552, *SE*			
	Objective 1	Objective 2	Other
Basic	15–37 odd, 47, 54	39–45 odd, 50	46, 55–74
Average	15–37 odd, 49, 54	39–45 odd, 50	55–74
Advanced	14–36 even, 49, 51–54	38–44 even, 50	55–71 (optional: 72–74)

Alternate Assignment _____

4 Assess

____ Open-Ended Assessment, *TWE*, p. 552
____ Enrichment, *CRM*, p. 602
____ Assessment, Mid-Chapter Test, *CRM*, p. 637
____ Assessment, Quiz, *CRM*, p. 636
____ algebra1.com/self_check_quiz

____ Closing the Gap for Absent Students, pp. 20–21

KEY *SE* = Student Edition   *TWE* = Teacher Wraparound Edition   *CRM* = Chapter Resource Masters

Graphing Calculator Investigation (p. 553)
A Follow-Up of Lesson 10-4

Teacher's Name _____ Dates _____

Grade _____ Class _____ M Tu W Th F

NCTM Standards
2, 6

Recommended Pacing	
Regular Average	Day 8 of 16
Regular Advanced	Day 8 of 16
Block Average	Day 5 of 9
Block Advanced	Day 4 of 8

Objectives
___ Use a graphing calculator to solve quadratic-linear systems.
___ State/local objectives: _____

Getting Started
Materials/Resources Needed _____

Teach
___ *Graphing Calculator and Spreadsheet Masters*, p. 67
___ algebra1.com/other_calculator_keystrokes

Assignment Guide, p. 553, *SE*	
All	1–6
Alternate Assignment	

Assess

KEY   *SE* = Student Edition   *TWE* = Teacher Wraparound Edition   *CRM* = Chapter Resource Masters

© Glencoe/McGraw-Hill   Glencoe Algebra 1

Lesson Planning Guide (pp. 554–560)

10-5

Teacher's Name _____ Dates _____

Grade _____ Class _____ M Tu W Th F

NCTM Standards
2, 6, 8, 9, 10

Recommended Pacing	
Regular Average	Days 9 & 10 of 16
Regular Advanced	Days 9 & 10 of 16
Block Average	Day 6 of 9
Block Advanced	Day 5 of 8

Objectives
___ Graph exponential functions.
___ Identify data that displays exponential behavior.
___ State/local objectives: _____

1 Focus
Materials/Resources Needed _____
_____
___ *5-Minute Check Transparencies*, Lesson 10-5
___ Mathematical Background, *TWE*, p. 522D
___ *TeacherWorks CD-ROM*

2 Teach
___ In-Class Examples, *TWE*, pp. 555–557
___ *Teaching Algebra with Manipulatives*, p. 180
___ *Interactive Chalkboard CD-ROM*, Lesson 10-5
___ algebra1.com/extra_examples
___ *Guide to Daily Intervention*, pp. 26–27
___ Daily Intervention, *TWE*, p. 558
___ Study Guide and Intervention, *CRM*, pp. 603–604
___ Reading to Learn Mathematics, *CRM*, p. 607
___ *TeacherWorks CD-ROM*

3 Practice/Apply
___ Skills Practice, *CRM*, p. 605
___ Practice, *CRM*, p. 606
___ Extra Practice, *SE*, p. 843
___ Differentiated Instruction, *TWE*, p. 557
___ *Graphing Calculator and Spreadsheet Masters*, pp. 42, 68
___ *Parent and Student Study Guide Workbook*, p. 79
___ Answer Key Transparencies, Lesson 10-5

Assignment Guide, pp. 558–560, *SE*			
	Objective 1	Objective 2	Other
Basic	13–21 odd, 25, 34	27–31 odd	33, 35, 42–64
Average	13–25 odd	27–31 odd	37–39, 42–64
Advanced	14–26 even	28–32 even	36–60 (optional: 61–64)
All	Practice Quiz 2 (1–5)		
Alternate Assignment			

4 Assess
___ Practice Quiz 2, *SE*, p. 560
___ Open-Ended Assessment, *TWE*, p. 560
___ Enrichment, *CRM*, p. 608
___ algebra1.com/self_check_quiz

___ *Closing the Gap for Absent Students*, pp. 20–21

KEY *SE* = Student Edition   *TWE* = Teacher Wraparound Edition   *CRM* = Chapter Resource Masters

Lesson Planning Guide (pp. 561–566)

10-6

Teacher's Name _____ Dates _____

Grade _____ Class _____ M Tu W Th F

NCTM Standards
2, 6, 8, 9, 10

Recommended Pacing	
Regular Average	Days 11 & 12 of 16
Regular Advanced	Days 11 & 12 of 16
Block Average	Day 7 of 9
Block Advanced	Day 6 of 8

Objectives
___ Solve problems involving exponential growth.
___ Solve problems involving exponential decay.
___ State/local objectives: _____

1 Focus
Materials/Resources Needed _____
_____
___ 5-Minute Check Transparencies, Lesson 10-6
___ Mathematical Background, *TWE*, p. 522D
___ TeacherWorks CD-ROM

2 Teach
___ In-Class Examples, *TWE*, pp. 562–563
___ *Interactive Chalkboard CD-ROM*, Lesson 10-6
___ algebra1.com/extra_examples
___ *Guide to Daily Intervention*, pp. 26–27
___ Study Guide and Intervention, *CRM*, pp. 609–610
___ Reading to Learn Mathematics, *CRM*, p. 613
___ *TeacherWorks CD-ROM*
___ Reading Mathematics, *SE*, p. 566

3 Practice/Apply
___ Skills Practice, *CRM*, p. 611
___ Practice, *CRM*, p. 612
___ Extra Practice, *SE*, p. 843
___ Differentiated Instruction, *TWE*, p. 562
___ Science and Mathematics Lab Manual, pp. 79–82
___ Parent and Student Study Guide Workbook, p. 80
___ Real-World Transparency and Master
___ Answer Key Transparencies, Lesson 10-6

Assignment Guide, pp. 563–565, *SE*			
	Objective 1	Objective 2	Other
Basic	9–12, 13, 15	17, 19	23, 24, 29–48
Average	9–15	17, 19	23, 24, 29–48
Advanced	11, 12, 14 18, 21, 22	16, 20, 25–28	23, 24, 29–45 (optional: 46–48)
Reading Mathematics	1–5		
Alternate Assignment			

4 Assess
___ Open-Ended Assessment, *TWE*, p. 565
___ Enrichment, *CRM*, p. 614
___ Assessment, Quiz, *CRM*, p. 636
___ algebra1.com/self_check_quiz

___ Closing the Gap for Absent Students, pp. 20–21

KEY *SE* = Student Edition *TWE* = Teacher Wraparound Edition *CRM* = Chapter Resource Masters

© Glencoe/McGraw-Hill Glencoe Algebra 1

Lesson Planning Guide (pp. 567–572)

10-7

Teacher's Name _____ Dates _____

Grade _____ Class _____ M Tu W Th F

NCTM Standards
1, 2, 3, 6, 8, 9, 10

Recommended Pacing	
Regular Average	Day 13 of 16
Regular Advanced	Day 13 of 16
Block Average	Day 8 of 9
Block Advanced	Day 7 of 8

Objectives
___ Recognize and extend geometric sequences.
___ Find geometric means.
___ State/local objectives: _____

---

1 Focus
Materials/Resources Needed _____
___ Building on Prior Knowledge, *TWE*, p. 567
___ *5-Minute Check Transparencies*, Lesson 10-7
___ Mathematical Background, *TWE*, p. 522D
___ *Prerequisite Skills Masters*, pp. 9–12, 47–48
___ *TeacherWorks CD-ROM*

2 Teach
___ In-Class Examples, *TWE*, pp. 568–570
___ *Teaching Algebra with Manipulatives*, p. 181
___ *Interactive Chalkboard CD-ROM*, Lesson 10-7
___ algebra1.com/extra_examples
___ *Guide to Daily Intervention*, pp. 26–27
___ Study Guide and Intervention, *CRM*, pp. 615–616
___ Reading to Learn Mathematics, *CRM*, p. 619
___ *TeacherWorks CD-ROM*

3 Practice/Apply
___ Skills Practice, *CRM*, p. 617
___ Practice, *CRM*, p. 618
___ Extra Practice, *SE*, p. 843
___ Differentiated Instruction, *TWE*, p. 568
___ *Graphing Calculator and Spreadsheet Masters*, p. 41
___ *Parent and Student Study Guide Workbook*, p. 81
___ *Answer Key Transparencies*, Lesson 10-7
___ algebra1.com/webquest

Assignment Guide, pp. 570–572, *SE*			
	Objective 1	Objective 2	Other
Basic	17–21 odd, 25–29 odd, 33–39 odd, 53, 61–63	43–53 odd	64, 65, 69–74
Average	17–41 odd, 55–57, 61–63	43–53 odd	64, 65, 69–74 (optional: 66–68)
Advanced	18–42 even, 56–62 even	44–54 even	57–61 odd, 63–74
Alternate Assignment			_____

4 Assess
___ Open-Ended Assessment, *TWE*, p. 572
___ Enrichment, *CRM*, p. 620
___ Assessment, Quiz, *CRM*, p. 572
___ algebra1.com/self_check_quiz

___ *Closing the Gap for Absent Students*, pp. 20–21

KEY *SE* = Student Edition *TWE* = Teacher Wraparound Edition *CRM* = Chapter Resource Masters

Algebra Activity (p. 573)
A Follow-Up of Lesson 10-7

10

Teacher's Name _____ Dates _____

Grade _____ Class _____ M Tu W Th F

NCTM Standards
1, 2, 6, 8, 9, 10

Recommended Pacing	
Regular Average	Day 14 of 16
Regular Advanced	Day 14 of 16
Block Average	Day 8 of 9
Block Advanced	Day 7 of 9

Objectives
___ Use a table to investigate rates of change.
___ State/local objectives: _____

Getting Started
Materials/Resources Needed *grid paper*

Teach
___ *Teaching Algebra with Manipulatives*, p. 182
___ *Glencoe Mathematics Classroom Manipulative Kit*
___ Teaching Strategy, *TWE*, p. 573
___ Graphing Calculator and Spreadsheet Masters, p. 69

Assignment Guide, p. 573, *SE*	
All	1–6
Alternate Assignment	

Assess
___ Study Notebook, *TWE*, p. 573

KEY *SE* = Student Edition   *TWE* = Teacher Wraparound Edition   *CRM* = Chapter Resource Masters

Review and Testing (pp. 574–581)

10

Teacher's Name _____ Dates _____

Grade _____ Class _____ M Tu W Th F

Recommended Pacing	
Regular Average	Days 15 & 16 of 16
Regular Advanced	Days 15 & 16 of 16
Block Average	Day 9 of 9
Block Advanced	Day 8 of 8

Assess

___ *Parent and Student Study Guide Workbook,* p. 82
___ Vocabulary and Concept Check, *SE,* p. 574
___ Vocabulary Test, *CRM,* p. 634
___ Lesson-by-Lesson Review, *SE,* pp. 574–578
___ Practice Test, *SE,* p. 579
___ Chapter 10 Tests, *CRM,* pp. 621–632
___ Open-Ended Assessment, *CRM,* p. 633
___ Standardized Test Practice, *SE,* pp. 580–581
___ Standardized Test Practice, *CRM,* pp. 639–640
___ Cumulative Review, *CRM,* p. 638
___ *Vocabulary PuzzleMaker CD-ROM*
___ algebra1.com/vocabulary_review
___ algebra1.com/chapter_test
___ algebra1.com/standardized_test
___ *MindJogger Videoquizzes VHS*
___ Unit 3 Test, *CRM,* pp. 641–642

Other Assessment Materials

- *TestCheck and Worksheet Builder CD-ROM*

KEY	*SE* = Student Edition	*TWE* = Teacher Wraparound Edition	*CRM* = Chapter Resource Masters

Lesson Planning Guide (pp. 586-592)

11-1

Teacher's Name _____ Dates _____

Grade _____ Class _____ M Tu W Th F

NCTM Standards
1, 2, 6, 8, 9, 10

Recommended Pacing	
Regular Average	Days 1 & 2 of 13
Regular Advanced	Days 1 & 2 of 14
Block Average	Days 1 & 2 of 7
Block Advanced	Days 1 & 2 of 8

Objectives
___ Simplify radical expressions using the Product Property of Square Roots.
___ Simplify radical expressions using the Quotient Property of Square Roots.
___ State/local objectives: _____

1 Focus
Materials/Resources Needed _____
___ Building on Prior Knowledge, *TWE*, p. 586
___ *5-Minute Check Transparencies*, Lesson 11-1
___ Mathematical Background, *TWE*, p. 584C
___ *TeacherWorks CD-ROM*

2 Teach
___ In-Class Examples, *TWE*, pp. 587–589
___ *Interactive Chalkboard CD-ROM*, Lesson 11-1
___ algebra1.com/extra_examples
___ algebra1.com/careers
___ *Guide to Daily Intervention*, pp. 28–29
___ Study Guide and Intervention, *CRM*, pp. 643–644
___ Reading to Learn Mathematics, *CRM*, p. 647
___ *TeacherWorks CD-ROM*

3 Practice/Apply
___ Skills Practice, *CRM*, p. 645
___ Practice, *CRM*, p. 646
___ Extra Practice, *SE*, p. 844
___ Differentiated Instruction, *TWE*, p. 588
___ *School-to-Career Masters*, p. 21
___ *Graphing Calculator and Spreadsheet Masters*, p. 70
___ *Parent and Student Study Guide Workbook*, p. 83
___ *Answer Key Transparencies*, Lesson 11-1
___ algebra1.com/webquest

Assignment Guide, pp. 589–592, *SE*			
	Objective 1	Objective 2	Other
Basic	15–25 odd	27–37 odd	39, 41–43, 50–53, 62–89
Average	15–25 odd, 39, 41	27–37 odd	45–47, 50–53, 62–89 (optional: 54–61)
Advanced	16–26 even	28–40 even	44–83 (optional: 84–89)

Alternate Assignment _____

4 Assess
___ Open-Ended Assessment, *TWE*, p. 592
___ Enrichment, *CRM*, p. 648
___ algebra1.com/self_check_quiz

___ *Closing the Gap for Absent Students*, pp. 22–23

KEY SE = Student Edition   TWE = Teacher Wraparound Edition   CRM = Chapter Resource Masters

© Glencoe/McGraw-Hill   Glencoe Algebra 1

Lesson Planning Guide (pp. 593-597)

11-2

Teacher's Name _____ Dates _____

Grade _____ Class _____ M Tu W Th F

NCTM Standards
1, 6, 8, 9, 10

Recommended Pacing	
Regular Average	Days 3 & 4 of 13
Regular Advanced	Days 3 & 4 of 14
Block Average	Days 2 & 3 of 7
Block Advanced	Days 2 & 3 of 8

Objectives
____ Add and subtract radical expressions.
____ Multiply radical expressions.
____ State/local objectives: _____

1 Focus
Materials/Resources Needed _____
____ *5-Minute Check Transparencies*, Lesson 11-2
____ *Mathematical Background, TWE*, p. 584C
____ *Prerequisite Skills Masters*, pp. 37–38
____ *TeacherWorks CD-ROM*

2 Teach
____ In-Class Examples, *TWE*, p. 594
____ *Teaching Algebra with Manipulatives*, p. 185
____ *Interactive Chalkboard CD-ROM*, Lesson 11-2
____ algebra1.com/extra_examples
____ algebra1.com/data_update
____ *Guide to Daily Intervention*, pp. 28–29
____ *Study Guide and Intervention, CRM*, pp. 649–650
____ *Reading to Learn Mathematics, CRM*, p. 653
____ *TeacherWorks CD-ROM*

3 Practice/Apply
____ *Skills Practice, CRM*, p. 651
____ *Practice, CRM*, p. 652
____ *Extra Practice, SE*, p. 844
____ *Differentiated Instruction, TWE*, p. 594
____ *Parent and Student Study Guide Workbook*, p. 84
____ *Answer Key Transparencies*, Lesson 11-2
____ *WebQuest and Projects Resources*, p. 61

Assignment Guide, pp. 595–597, *SE*			
	Objective 1	Objective 2	Other
Basic	15–25 odd	31, 33	39, 41, 42, 49–76
Average	15–29 odd	31–37 odd	39, 41–44, 49–76
Advanced	14–28 even	30–36 even	38, 40, 45–70 (optional: 71–76)

Alternate Assignment _____

4 Assess
____ Open-Ended Assessment, *TWE*, p. 597
____ Enrichment, *CRM*, p. 654
____ Assessment, Quiz, *CRM*, p. 699
____ algebra1.com/self_check_quiz

____ *Closing the Gap for Absent Students*, pp. 22–23

KEY *SE* = Student Edition   *TWE* = Teacher Wraparound Edition   *CRM* = Chapter Resource Masters

© Glencoe/McGraw-Hill   Glencoe Algebra 1

Lesson Planning Guide (pp. 598–603)

11-3

Teacher's Name _____ Dates _____

Grade _____ Class _____ M Tu W Th F

NCTM Standards
2, 6, 8, 9, 10

Recommended Pacing	
Regular Average	Day 5 of 13
Regular Advanced	Days 5 & 6 of 14
Block Average	Day 3 of 7
Block Advanced	Days 3 & 4 of 8

Objectives

____ Solve radical equations.
____ Solve radical equations with extraneous solutions.
____ State/local objectives: _____

1 Focus

Materials/Resources Needed _____
_____

____ Building on Prior Knowledge, *TWE*, p. 599
____ *5-Minute Check Transparencies*, Lesson 11-3
____ Mathematical Background, *TWE*, p. 584D
____ *TeacherWorks CD-ROM*

2 Teach

____ In-Class Examples, *TWE*, p. 599
____ *Interactive Chalkboard CD-ROM*, Lesson 11-3
____ algebra1.com/extra_examples
____ *Guide to Daily Intervention*, pp. 28–29
____ Daily Intervention, *TWE*, p. 600
____ Study Guide and Intervention, *CRM*, pp. 655–656
____ Reading to Learn Mathematics, *CRM*, p. 659
____ *TeacherWorks CD-ROM*
____ *Multimedia Applications Masters*

3 Practice/Apply

____ Skills Practice, *CRM*, p. 657
____ Practice, *CRM*, p. 658
____ Extra Practice, *SE*, p. 844
____ Differentiated Instruction, *TWE*, p. 599
____ *Graphing Calculator and Spreadsheet Masters*, pp. 43, 71
____ *Parent and Student Study Guide Workbook*, p. 85
____ *Answer Key Transparencies*, Lesson 11-3
____ *WebQuest and Projects Resources*, p. 61

Assignment Guide, pp. 601–603, *SE*			
	Objective 1	Objective 2	Other
Basic	17–33 odd	35–47 odd	48, 49, 60–63, 70–89
Average	17–33 odd	35–47 odd	48–53, 60–63, 70–89 (optional: 64–69)
Advanced	18–34 even	36–46 even	50–85 (optional: 86–89)
All	Practice Quiz 1 (1–10)		
Alternate Assignment	_____		

4 Assess

____ Practice Quiz 1, *SE*, p. 603
____ Open-Ended Assessment, *TWE*, p. 603
____ Enrichment, *CRM*, p. 660
____ algebra1.com/self_check_quiz

____ Closing the Gap for Absent Students, pp. 22–23

KEY	*SE* = Student Edition	*TWE* = Teacher Wraparound Edition	*CRM* = Chapter Resource Masters

© Glencoe/McGraw-Hill — Glencoe Algebra 1

Graphing Calculator Investigation (p. 604)
A Follow-Up of Lesson 11-3

11

Teacher's Name _____ Dates _____

Grade _____ Class _____ M Tu W Th F

NCTM Standards
2, 6, 8

Recommended Pacing	
Regular Average	Day 6 of 13
Regular Advanced	Day 7 of 14
Block Average	Day 4 of 7
Block Advanced	Day 5 of 8

Objectives
___ Use a graphing calculator to graph radical equations.
___ State/local objectives: _____

Getting Started
Materials/Resources Needed _____
_____
_____

Teach
___ *Graphing Calculator and Spreadsheet Masters*, p. 72
___ algebra1.com/other_calculator_keystrokes

Assignment Guide, p. 604, *SE*	
All	1–12
Alternate Assignment	

Assess

KEY   *SE* = Student Edition   *TWE* = Teacher Wraparound Edition   *CRM* = Chapter Resource Masters

© Glencoe/McGraw-Hill   Glencoe Algebra 1

Lesson Planning Guide (pp. 605–610)

11-4

Teacher's Name _____ Dates _____

Grade _____ Class _____ M Tu W Th F

NCTM Standards
1, 2, 3, 6, 8, 9, 10

Recommended Pacing	
Regular Average	Day 7 of 13
Regular Advanced	Days 8 of 14
Block Average	Day 4 of 7
Block Advanced	Day 5 of 8

Objectives
___ Solve problems by using the Pythagorean Theorem.
___ Determine whether a triangle is a right triangle.
___ State/local objectives: _____

1 Focus
Materials/Resources Needed _____

___ *5-Minute Check Transparencies*, Lesson 11-4
___ *Mathematical Background, TWE*, p. 584D
___ *TeacherWorks CD-ROM*

2 Teach
___ In-Class Examples, *TWE*, pp. 606–607
___ *Teaching Algebra with Manipulatives*, pp. 186–189
___ *Interactive Chalkboard CD-ROM*, Lesson 11-4
___ algebra1.com/extra_examples
___ *Guide to Daily Intervention*, pp. 28–29
___ *Study Guide and Intervention, CRM*, pp. 661–662
___ *Reading to Learn Mathematics, CRM*, p. 665
___ *TeacherWorks CD-ROM*

3 Practice/Apply
___ *Skills Practice, CRM*, p. 663
___ *Practice, CRM*, p. 664
___ *Extra Practice, SE*, p. 845
___ *Differentiated Instruction, TWE*, p. 607
___ *School-to-Career Masters*, p. 22
___ *Graphing Calculator and Spreadsheet Masters*, p. 44
___ *Parent and Student Study Guide Workbook*, p. 86
___ *Real-World Transparency and Master*
___ *Answer Key Transparencies*, Lesson 11-4
___ *AlgePASS CD-ROM*, Lesson 32
___ *WebQuest and Projects Resources*, p. 62

Assignment Guide, pp. 608–610, *SE*			
	Objective 1	Objective 2	Other
Basic	13–29, odd, 37	31–35 odd	41–43, 49–68
Average	13–29 odd, 37, 39	31–35 odd	41–43, 45, 48–68
Advanced	14–30 even, 38, 40	32–36 even	44, 46–62 (optional: 63–68)

Alternate Assignment _____

4 Assess
___ Open-Ended Assessment, *TWE*, p. 610
___ *Enrichment, CRM*, p. 666
___ Assessment, Mid–Chapter Test, *CRM*, p. 701
___ Assessment, Quiz, *CRM*, p. 669
___ algebra1.com/self_check_quiz

___ *Closing the Gap for Absent Students*, pp. 22–23

KEY   *SE* = Student Edition   *TWE* = Teacher Wraparound Edition   *CRM* = Chapter Resource Masters

© Glencoe/McGraw-Hill   Glencoe Algebra 1

Lesson Planning Guide (pp. 611–615)

11-5

Teacher's Name _____ Dates _____

Grade _____ Class _____ M Tu W Th F

NCTM Standards
1, 2, 3, 6, 8, 9, 10

Recommended Pacing	
Regular Average	Day 8 of 13
Regular Advanced	Day 9 of 14
Block Average	Day 5 of 7
Block Advanced	Day 6 of 8

Objectives

____ Find the distance between two points on the coordinate plane.
____ Find a point that is a given distance from a second point in a plane.
____ State/local objectives: _____

1 Focus

Materials/Resources Needed _____

____ *5-Minute Check Transparencies*, Lesson 11-5
____ Mathematical Background, *TWE*, p. 584D
____ *TeacherWorks CD-ROM*

2 Teach

____ In-Class Examples, *TWE*, p. 612
____ *Interactive Chalkboard CD-ROM*, Lesson 11-5
____ algebra1.com/extra_examples
____ *Guide to Daily Intervention*, pp. 28–29
____ Daily Intervention, *TWE*, p. 612
____ Study Guide and Intervention, *CRM*, pp. 667–668
____ Reading to Learn Mathematics, *CRM*, p. 671
____ *TeacherWorks CD-ROM*

3 Practice/Apply

____ Skills Practice, *CRM*, p. 669
____ Practice, *CRM*, p. 670
____ Extra Practice, *SE*, p. 845
____ Differentiated Instruction, *TWE*, p. 613
____ *Science and Mathematics Lab Manual*, pp. 87–92
____ *Parent and Student Study Guide Workbook*, p. 87
____ *Answer Key Transparencies*, Lesson 11-5
____ *WebQuest and Projects Resources*, p. 62

Assignment Guide, pp. 613–615, *SE*			
	Objective 1	Objective 2	Other
Basic	13–25 odd, 37–39	27–31 odd	43–68
Average	13–25 odd, 33, 37–39	27–31 odd	43–68
Advanced	14–26 even, 34, 40–42	28–32 even, 36	43–62 (optional: 63–68)

Alternate Assignment _____

4 Assess

____ Open-Ended Assessment, *TWE*, p. 615
____ Enrichment, *CRM*, p. 672
____ algebra1.com/self_check_quiz

____ Closing the Gap for Absent Students, pp. 22–23

KEY *SE* = Student Edition   *TWE* = Teacher Wraparound Edition   *CRM* = Chapter Resource Masters

© Glencoe/McGraw-Hill   *Glencoe Algebra 1*

Lesson Planning Guide (pp. 616–621)

11-6

Teacher's Name _____ Dates _____

Grade _____ Class _____ M Tu W Th F

NCTM Standards
1, 2, 3, 6, 8, 9, 10

Recommended Pacing	
Regular Average	Day 9 of 13
Regular Advanced	Day 10 of 14
Block Average	Day 5 of 7
Block Advanced	Day 6 of 8

Objectives
___ Determine whether two triangles are similar.
___ Find the unknown measures of sides of two similar triangles.
___ State/local objectives: _____

1 Focus
Materials/Resources Needed _____

___ Building on Prior Knowledge, *TWE*, p. 616
___ 5-Minute Check Transparencies, Lesson 11-6
___ Mathematical Background, *TWE*, p. 584D
___ Prerequisite Skills Masters, pp. 61–62
___ TeacherWorks CD-ROM

2 Teach
___ In-Class Examples, *TWE*, pp. 617–618
___ Teaching Algebra with Manipulatives, p. 190
___ Interactive Chalkboard CD-ROM, Lesson 11-6
___ algebra1.com/extra_examples
___ Guide to Daily Intervention, pp. 28–29
___ Daily Intervention, *TWE*, p. 618
___ Study Guide and Intervention, *CRM*, pp. 673–674
___ Reading to Learn Mathematics, *CRM*, p. 677
___ TeacherWorks CD-ROM

3 Practice/Apply
___ Skills Practice, *CRM*, p. 675
___ Practice, *CRM*, p. 676
___ Extra Practice, *SE*, p. 845
___ Differentiated Instruction, *TWE*, p. 618
___ Parent and Student Study Guide Workbook, p. 88
___ Answer Key Transparencies, Lesson 11-6

Assignment Guide, pp. 618–621, *SE*			
	Objective 1	Objective 2	Other
Basic	11–15 odd	17–27 odd	33–61
Average	11–15 odd	17–27 odd, 29, 30	33–61
Advanced	12–16 even	18–28 even, 29–32	33–55 (optional: 56–61)
All		Practice Quiz 2 (1–10)	
Alternate Assignment			

4 Assess
___ Practice Quiz 2, *SE*, p. 621
___ Open-Ended Assessment, *TWE*, p. 621
___ Enrichment, *CRM*, p. 678

___ Assessment, Quiz, *CRM*, p. 700
___ algebra1.com/self_check_quiz

___ Closing the Gap for Absent Students, pp. 22–23

KEY *SE* = Student Edition   *TWE* = Teacher Wraparound Edition   *CRM* = Chapter Resource Masters

© Glencoe/McGraw-Hill   Glencoe Algebra 1

Algebra Activity (p. 622)
A Preview of Lesson 11-7

11

Teacher's Name _____ Dates _____

Grade _____ Class _____ M Tu W Th F

NCTM Standards
1, 3, 7

Recommended Pacing	
Regular Average	Day 10 of 13
Regular Advanced	Day 11 of 14
Block Average	Day 6 of 7
Block Advanced	Day 7 of 8

Objectives
___ Use paper triangles to investigate trigonometric ratios.
___ State/local objectives: _____

Getting Started
Materials/Resources Needed ruler, grid paper, protractor
_____
_____

Teach
___ *Teaching Algebra with Manipulatives*, p. 191
___ *Glencoe Mathematics Classroom Manipulative Kit*
___ Teaching Strategy, *TWE*, p. 622

Assignment Guide, p. 622, *SE*	
All	1–3
Alternate Assignment	

Assess

KEY   *SE* = Student Edition   *TWE* = Teacher Wraparound Edition   *CRM* = Chapter Resource Masters

© Glencoe/McGraw-Hill   Glencoe Algebra 1

Lesson Planning Guide (pp. 623–631)

11-7

Teacher's Name _____ Dates _____

Grade _____ Class _____ M Tu W Th F

NCTM Standards
3, 6, 8, 9, 10

Recommended Pacing	
Regular Average	Day 11 of 13
Regular Advanced	Day 12 of 14
Block Average	Day 6 of 7
Block Advanced	Day 7 of 8

Objectives
___ Define the sine, cosine, and tangent ratios.
___ Use trigonometric ratios to solve right triangles.
___ State/local objectives: _____

1 Focus
Materials/Resources Needed _____
_____
___ *5-Minute Check Transparencies*, Lesson 11-7
___ Mathematical Background, *TWE*, p. 584D
___ *TeacherWorks CD-ROM*

2 Teach
___ In-Class Examples, *TWE*, pp. 624–626
___ *Teaching Algebra with Manipulatives*, p. 192
___ *Interactive Chalkboard CD-ROM*, Lesson 11-7
___ algebra1.com/extra_examples
___ *Guide to Daily Intervention*, pp. 28–29
___ Study Guide and Intervention, *CRM*, pp. 679–680
___ Reading to Learn Mathematics, *CRM*, p. 683
___ *TeacherWorks CD-ROM*
___ Reading Mathematics, *SE*, p. 631

3 Practice/Apply
___ Skills Practice, *CRM*, p. 681
___ Practice, *CRM*, p. 682
___ Extra Practice, *SE*, p. 846
___ Differentiated Instruction, *TWE*, p. 624
___ Parent and Student Study Guide Workbook, p. 89
___ *Answer Key Transparencies*, Lesson 11-7

Assignment Guide, pp. 627–630, *SE*			
	Objective 1	Objective 2	Other
Basic	19–49 odd	53–59 odd, 61	62, 66–79
Average	19–51 odd	53–59 odd, 61–64	66–79
Advanced	20–50 even	52–60 even, 63–65	66–79
Reading Mathematics	1–3		
Alternate Assignment			

4 Assess
___ Open-Ended Assessment, *TWE*, p. 630
___ Enrichment, *CRM*, p. 684
___ Assessment, Quiz, *CRM*, p. 700
___ algebra1.com/self_check_quiz

___ *Closing the Gap for Absent Students*, pp. 22–23

KEY *SE* = Student Edition   *TWE* = Teacher Wraparound Edition   *CRM* = Chapter Resource Masters

© Glencoe/McGraw-Hill   *Glencoe Algebra 1*

11 Review and Testing (pp. 632–639)

Teacher's Name _____   Dates _____

Grade _____   Class _____   M Tu W Th F

Recommended Pacing	
Regular Average	Days 12 & 13 of 13
Regular Advanced	Days 13 & 14 of 14
Block Average	Day 7 of 7
Block Advanced	Day 8 of 8

Assess

____ *Parent and Student Study Guide Workbook*, p. 90
____ Vocabulary and Concept Check, *SE*, p. 632
____ Vocabulary Test, *CRM*, p. 698
____ Lesson-by-Lesson Review, *SE*, pp. 632–636
____ Practice Test, *SE*, p. 637
____ Chapter 11 Tests, *CRM*, pp. 685–696
____ Open-Ended Assessment, *CRM*, p. 697
____ Standardized Test Practice, *SE*, pp. 638–639
____ Standardized Test Practice, *CRM*, pp. 703–704
____ Cumulative Review, *CRM*, p. 702
____ *Vocabulary PuzzleMaker CD-ROM*
____ algebra1.com/vocabulary_review
____ algebra1.com/chapter_test
____ algebra1.com/standardized_test
____ *MindJogger Videoquizzes VHS*

Other Assessment Materials

- *TestCheck and Worksheet Builder CD-ROM*

KEY   *SE* = Student Edition   *TWE* = Teacher Wraparound Edition   *CRM* = Chapter Resource Masters

Lesson Planning Guide (pp. 642–647)

12-1

Teacher's Name _____ Dates _____

Grade _____ Class _____ M Tu W Th F

NCTM Standards
1, 2, 6, 8, 9, 10

Recommended Pacing	
Regular Average	Day 1 of 14
Regular Advanced	Day 1 of 17
Block Average	Day 1 of 7
Block Advanced	Day 1 of 9

Objectives
____ Graph inverse variations.
____ Solve problems involving inverse variation.
____ State/local objectives: _____

1 Focus
Materials/Resources Needed _____

____ Building on Prior Knowledge, *TWE*, p. 642
____ 5-Minute Check Transparencies, Lesson 12-1
____ Mathematical Background, *TWE*, p. 640C
____ TeacherWorks CD-ROM

2 Teach
____ In-Class Examples, *TWE*, pp. 643–644
____ *Interactive Chalkboard CD-ROM*, Lesson 12-1
____ algebra1.com/extra_examples
____ *Guide to Daily Intervention*, pp. 30–31
____ Study Guide and Intervention, *CRM*, pp. 705–706
____ Reading to Learn Mathematics, *CRM*, p. 709
____ TeacherWorks CD-ROM

3 Practice/Apply
____ Skills Practice, *CRM*, p. 707
____ Practice, *CRM*, p. 708
____ Extra Practice, *SE*, p. 846
____ Differentiated Instruction, *TWE*, p. 644
____ *Parent and Student Study Guide Workbook*, p. 91
____ Answer Key Transparencies, Lesson 12-1

Assignment Guide, pp. 645–647, *SE*			
	Objective 1	Objective 2	Other
Basic	11–15 odd	17–23 odd, 31–33	38–60
Average	11–15 odd	17–31 odd, 34–37	38–60
Advanced	12–16 even	18–30 even, 34–37	38–54 (optional: 55–60)
Alternate Assignment			

4 Assess
____ Open-Ended Assessment, *TWE*, p. 647
____ Enrichment, *CRM*, p. 710
____ algebra1.com/self_check_quiz

____ *Closing the Gap for Absent Students*, pp. 24–25

KEY *SE* = Student Edition   *TWE* = Teacher Wraparound Edition   *CRM* = Chapter Resource Masters

© Glencoe/McGraw-Hill   *Glencoe Algebra 1*

Lesson Planning Guide (pp. 648–653)

12-2

Teacher's Name _____ Dates _____
Grade _____ Class _____ M Tu W Th F

NCTM Standards
1, 2, 6, 8, 9, 10

Recommended Pacing	
Regular Average	Day 2 of 14
Regular Advanced	Day 2 of 17
Block Average	Day 1 of 7
Block Advanced	Day 1 of 9

Objectives
___ Identify values excluded from the domain of a rational expression.
___ Simplify rational expressions.
___ State/local objectives: _____

1 Focus
Materials/Resources Needed _____
_____
___ Building on Prior Knowledge, *TWE*, p. 648
___ *5-Minute Check Transparencies*, Lesson 12-2
___ Mathematical Background, *TWE*, p. 640C
___ *TeacherWorks CD-ROM*

2 Teach
___ In-Class Examples, *TWE*, pp. 649–650
___ *Interactive Chalkboard CD-ROM*, Lesson 12-2
___ algebra1.com/extra_examples
___ algebra1.com/careers
___ *Guide to Daily Intervention*, pp. 30–31
___ Daily Intervention, *TWE*, p. 649
___ Study Guide and Intervention, *CRM*, pp. 711–712
___ Reading to Learn Mathematics, *CRM*, p. 715
___ *TeacherWorks CD-ROM*

3 Practice/Apply
___ Skills Practice, *CRM*, p. 713
___ Practice, *CRM*, p. 714
___ Extra Practice, *SE*, p. 846
___ Differentiated Instruction, *TWE*, p. 650
___ *Science and Mathematics Lab Manual*, pp. 83–86
___ *Parent and Student Study Guide Workbook*, p. 92
___ *Real-World Transparency and Master*
___ *Answer Key Transparencies*, Lesson 12-2
___ *AlgePASS CD-ROM*, Lesson 12-2
___ algebra1.com/webquest

Assignment Guide, pp. 651–653, *SE*			
	Objective 1	Objective 2	Other
Basic	17–23 odd	25–41 odd	42–48, 55–80
Average	17–23 odd	25–41 odd	46–52, 55–80
Advanced	16–22 even	24–40 even	49–74 (optional: 75–80)

Alternate Assignment _____

4 Assess
___ Open-Ended Assessment, *TWE*, p. 653
___ Enrichment, *CRM*, p. 716
___ algebra1.com/self_check_quiz

___ Closing the Gap for Absent Students, pp. 24–25

KEY *SE* = Student Edition   *TWE* = Teacher Wraparound Edition   *CRM* = Chapter Resource Masters

© Glencoe/McGraw-Hill   Glencoe Algebra 1

12 Graphing Calculator Investigation (p. 654)
A Follow-Up of Lesson 12-2

Teacher's Name _____ Dates _____

Grade _____ Class _____ M Tu W Th F

NCTM Standards
2, 8

Recommended Pacing	
Regular Average	Day 2 of 14
Regular Advanced	Day 3 of 17
Block Average	Day 1 of 7
Block Advanced	Day 2 of 9

Objectives
___ Use a graphing calculator to check simplified rational expressions.
___ State/local objectives: _____

Getting Started
Materials/Resources Needed _____
_____
_____

Teach
___ *Graphing Calculator and Spreadsheet Masters*, p. 73
___ algebra1.com/other_calculator_keystrokes

Assignment Guide, p. 654, *SE*	
All	1–4
Alternate Assignment	

Assess

KEY   *SE* = Student Edition   *TWE* = Teacher Wraparound Edition   *CRM* = Chapter Resource Masters

© Glencoe/McGraw-Hill   Glencoe Algebra 1

Lesson Planning Guide (pp. 655–659)

12-3

Teacher's Name _____ Dates _____

Grade _____ Class _____ M Tu W Th F

NCTM Standards
1, 2, 6, 8, 9, 10

Recommended Pacing	
Regular Average	Day 3 of 14
Regular Advanced	Day 4 of 17
Block Average	Day 2 of 7
Block Advanced	Day 3 of 9

Objectives
___ Multiply rational expressions.
___ Use dimensional analysis with multiplication.
___ State/local objectives: _____

1 Focus
Materials/Resources Needed _____
_____
___ Building on Prior Knowledge, *TWE*, p. 655
___ 5-Minute Check Transparencies, Lesson 12-3
___ Mathematical Background, *TWE*, p. 640C
___ *TeacherWorks CD-ROM*

2 Teach
___ In-Class Examples, *TWE*, p. 656
___ *Interactive Chalkboard CD-ROM*, Lesson 12-3
___ algebra1.com/extra_examples
___ *Guide to Daily Intervention*, pp. 30–31
___ Daily Intervention, *TWE*, p. 657
___ Study Guide and Intervention, *CRM*, pp. 717–718
___ Reading to Learn Mathematics, *CRM*, p. 721
___ *TeacherWorks CD-ROM*

3 Practice/Apply
___ Skills Practice, *CRM*, p. 719
___ Practice, *CRM*, p. 720
___ Extra Practice, *SE*, p. 847
___ Differentiated Instruction, *TWE*, p. 656
___ Parent and Student Study Guide Workbook, p. 93
___ Answer Key Transparencies, Lesson 12-3

Assignment Guide, pp. 657–659, *SE*			
	Objective 1	Objective 2	Other
Basic	13–27 odd	29–31 odd	35–61
Average	13–27 odd	29–31 odd	33, 35–61
Advanced	12–26 even	28, 30	32, 34–55 (optional: 56–61)
All	Practice Quiz 1 (1–10)		
Alternate Assignment			

4 Assess
___ Practice Quiz 1, *SE*, p. 659
___ Open-Ended Assessment, *TWE*, p. 659
___ Enrichment, *CRM*, p. 722

___ Assessment, Quiz, *CRM*, p. 773
___ algebra1.com/self_check_quiz

___ Closing the Gap for Absent Students, pp. 24–25

KEY   *SE* = Student Edition   *TWE* = Teacher Wraparound Edition   *CRM* = Chapter Resource Masters

© Glencoe/McGraw-Hill   Glencoe Algebra 1

Lesson Planning Guide (pp. 660–665)

12-4

Teacher's Name _____ Dates _____

Grade _____ Class _____ M Tu W Th F

NCTM Standards
1, 2, 6, 8, 9, 10

Recommended Pacing	
Regular Average	Day 4 of 14
Regular Advanced	Days 5 & 6 of 17
Block Average	Day 2 of 7
Block Advanced	Days 3 & 4 of 9

Objectives
___ Divide rational expressions.
___ Use dimensional analysis with division.
___ State/local objectives: _____

1 Focus
Materials/Resources Needed _____
_____

___ Building on Prior Knowledge, *TWE*, p. 660
___ *5-Minute Check Transparencies*, Lesson 12-4
___ Mathematical Background, *TWE*, p. 640D
___ *TeacherWorks CD-ROM*

2 Teach
___ In-Class Examples, *TWE*, p. 661
___ *Interactive Chalkboard CD-ROM*, Lesson 12-4
___ algebra1.com/extra_examples
___ *Guide to Daily Intervention*, pp. 30–31
___ Study Guide and Intervention, *CRM*, pp. 723–724
___ Reading to Learn Mathematics, *CRM*, p. 727
___ *TeacherWorks CD-ROM*
___ Reading Mathematics, *SE*, p. 665

3 Practice/Apply
___ Skills Practice, *CRM*, p. 725
___ Practice, *CRM*, p. 726
___ Extra Practice, *SE*, p. 847
___ Differentiated Instruction, *TWE*, p. 661
___ *Graphing Calculator and Spreadsheet Masters*, p. 46
___ *Parent and Student Study Guide Workbook*, p. 94
___ *Answer Key Transparencies*, Lesson 12-4

Assignment Guide, pp. 662–664, *SE*			
	Objective 1	Objective 2	Other
Basic	13–23 odd, 29, 31, 33	25, 27	35, 37–39, 42, 45–75
Average	13–23 odd, 29, 31, 33	25, 27	35, 37, 40–42, 45–75
Advanced	14–24 even, 30, 32, 34	26, 28	36, 42, 45–69 (optional: 70–75)
Reading Mathematics	1–12		
Alternate Assignment			

4 Assess
___ Open-Ended Assessment, *TWE*, p. 664
___ Enrichment, *CRM*, p. 728
___ algebra1.com/self_check_quiz

___ Closing the Gap for Absent Students, pp. 24–25

KEY *SE* = Student Edition   *TWE* = Teacher Wraparound Edition   *CRM* = Chapter Resource Masters

© Glencoe/McGraw-Hill   *Glencoe Algebra 1*

Lesson Planning Guide (pp. 666–671)

12-5

Teacher's Name _____ Dates _____
Grade _____ Class _____ M Tu W Th F

NCTM Standards
1, 2, 6, 8, 9, 10

Recommended Pacing	
Regular Average	Day 5 of 14
Regular Advanced	Days 7 & 8 of 17
Block Average	Day 3 of 7
Block Advanced	Days 4 & 5 of 9

Objectives
___ Divide a polynomial by a monomial.
___ Divide a polynomial by a binomial.
___ State/local objectives: _____

1 Focus
Materials/Resources Needed _____

___ Building on Prior Knowledge, *TWE*, p. 666
___ *5-Minute Check Transparencies*, Lesson 12-5
___ Mathematical Background, *TWE*, p. 640D
___ *TeacherWorks CD-ROM*

2 Teach
___ In-Class Examples, *TWE*, pp. 667–668
___ *Teaching Algebra with Manipulatives*, pp. 195–197
___ *Interactive Chalkboard CD-ROM*, Lesson 12-5
___ algebra1.com/extra_examples
___ *Guide to Daily Intervention*, pp. 30–31
___ Study Guide and Intervention, *CRM*, pp. 729–730
___ Reading to Learn Mathematics, *CRM*, p. 733
___ *TeacherWorks CD-ROM*

3 Practice/Apply
___ Skills Practice, *CRM*, p. 731
___ Practice, *CRM*, p. 732
___ Extra Practice, *SE*, p. 847
___ Differentiated Instruction, *TWE*, p. 668
___ *Parent and Student Study Guide Workbook*, p. 95
___ *Answer Key Transparencies*, Lesson 12-5

Assignment Guide, pp. 669–671, *SE*			
	Objective 1	Objective 2	Other
Basic	11, 13	15–29 odd	31–33, 40–60
Average	11, 13	15–29 odd	33–35, 39–60
Advanced	12, 14	16–30 even	36–56 (optional: 57–60)

Alternate Assignment _____

4 Assess
___ Open-Ended Assessment, *TWE*, p. 671
___ Enrichment, *CRM*, p. 734
___ Assessment, Mid-Chapter Test, *CRM*, p. 775
___ Assessment, Quiz, *CRM*, p. 773
___ algebra1.com/self_check_quiz

___ Closing the Gap for Absent Students, pp. 24–25

KEY   *SE* = Student Edition   *TWE* = Teacher Wraparound Edition   *CRM* = Chapter Resource Masters

© Glencoe/McGraw-Hill   Glencoe Algebra 1

Lesson Planning Guide (pp. 672–677)

12-6

Teacher's Name _____ Dates _____

Grade _____ Class _____ M Tu W Th F

NCTM Standards
1, 2, 6, 8, 9, 10

Recommended Pacing	
Regular Average	Day 6 of 14
Regular Advanced	Day 9 of 17
Block Average	Day 3 of 7
Block Advanced	Day 5 of 9

Objectives
____ Add rational expressions with like denominators.
____ Subtract rational expressions with like denominators.
____ State/local objectives: _____

1 Focus
Materials/Resources Needed _____
_____
____ Building on Prior Knowledge, *TWE*, p. 672
____ *5-Minute Check Transparencies*, Lesson 12-6
____ Mathematical Background, *TWE*, p. 640D
____ *TeacherWorks CD-ROM*

2 Teach
____ In-Class Examples, *TWE*, pp. 673–674
____ *Interactive Chalkboard CD-ROM*, Lesson 12-6
____ algebra1.com/extra_examples
____ *Guide to Daily Intervention*, pp. 30–31
____ Daily Intervention, *TWE*, p. 674
____ Study Guide and Intervention, *CRM*, pp. 735–736
____ Reading to Learn Mathematics, *CRM*, p. 739
____ *TeacherWorks CD-ROM*

3 Practice/Apply
____ Skills Practice, *CRM*, p. 737
____ Practice, *CRM*, p. 738
____ Extra Practice, *SE*, p. 848
____ Differentiated Instruction, *TWE*, p. 674
____ *Parent and Student Study Guide Workbook*, p. 96
____ *Answer Key Transparencies*, Lesson 12-6

Assignment Guide, pp. 674–677, *SE*			
	Objective 1	Objective 2	Other
Basic	15–27 odd	29–41 odd	43, 45, 46, 49–71
Average	15–27 odd	29–41 odd	43, 45–71
Advanced	14–26 even	28–40 even	42, 44–62 (optional: 63–71)
All	Practice Quiz 2 (1–10)		
Alternate Assignment	_____		

4 Assess
____ Practice Quiz 2, *SE*, p. 677
____ Open-Ended Assessment, *TWE*, p. 677
____ Enrichment, *CRM*, p. 740
____ algebra1.com/self_check_quiz

____ Closing the Gap for Absent Students, pp. 24–25

KEY *SE* = Student Edition   *TWE* = Teacher Wraparound Edition   *CRM* = Chapter Resource Masters

© Glencoe/McGraw-Hill   Glencoe Algebra 1

Lesson Planning Guide (pp. 678–683)

12-7

Teacher's Name _____ Dates _____

Grade _____ Class _____ M Tu W Th F

NCTM Standards
1, 2, 6, 8, 9, 10

Recommended Pacing	
Regular Average	Days 7 & 8 of 14
Regular Advanced	Days 10 & 11 of 17
Block Average	Day 4 of 7
Block Advanced	Day 6 of 9

Objectives
____ Add rational expressions with unlike denominators.
____ Subtract rational expressions with unlike denominators.
____ State/local objectives: _____

---

1 Focus
Materials/Resources Needed _____

____ Building on Prior Knowledge, *TWE*, p. 679
____ *5-Minute Check Transparencies*, Lesson 12-7
____ Mathematical Background, *TWE*, p. 640D
____ *Prerequisite Skills Masters*, pp. 17–18
____ *TeacherWorks CD-ROM*

2 Teach
____ In-Class Examples, *TWE*, pp. 679–680
____ *Interactive Chalkboard CD-ROM*, Lesson 12-7
____ algebra1.com/extra_examples
____ *Guide to Daily Intervention*, pp. 30–31
____ Study Guide and Intervention, *CRM*, pp. 741–742
____ Reading to Learn Mathematics, *CRM*, p. 745
____ *TeacherWorks CD-ROM*

3 Practice/Apply
____ Skills Practice, *CRM*, p. 743
____ Practice, *CRM*, p. 744
____ Extra Practice, *SE*, p. 848
____ Differentiated Instruction, *TWE*, p. 681
____ *School-to-Career Masters*, p. 23
____ *Parent and Student Study Guide Workbook*, p. 97
____ *Answer Key Transparencies*, Lesson 12-7

Assignment Guide, pp. 681–683, *SE*			
	Objective 1	Objective 2	Other
Basic	23–37 odd	39–53 odd	17–21 odd, 55, 57–77
Average	23–37 odd	39–53 odd	17–21 odd, 55, 57–77
Advanced	22–36 even	38–52 even	16–20 even, 54, 56, 58–71, (optional: 72–77)
Alternate Assignment			

4 Assess
____ Open-Ended Assessment, *TWE*, p. 683
____ Enrichment, *CRM*, p. 746
____ Assessment, Quiz, *CRM*, p. 774
____ algebra1.com/self_check_quiz

____ *Closing the Gap for Absent Students*, pp. 24–25

KEY *SE* = Student Edition *TWE* = Teacher Wraparound Edition *CRM* = Chapter Resource Masters

© Glencoe/McGraw-Hill   Glencoe Algebra 1

Lesson Planning Guide (pp. 684–689)

12-8

Teacher's Name _____ Dates _____

Grade _____ Class _____ M Tu W Th F

NCTM Standards
1, 2, 6, 8, 9, 10

Recommended Pacing	
Regular Average	Days 9 & 10 of 14
Regular Advanced	Days 12 & 13 of 17
Block Average	Day 5 of 7
Block Advanced	Day 7 of 9

Objectives
___ Simplify mixed expressions.
___ Simplify complex fractions.
___ State/local objectives: _____

1 Focus
Materials/Resources Needed _____
_____
___ *5-Minute Check Transparencies*, Lesson 12-8
___ Mathematical Background, *TWE*, p. 640D
___ *TeacherWorks CD-ROM*

2 Teach
___ In-Class Examples, *TWE*, pp. 685–686
___ *Interactive Chalkboard CD-ROM*, Lesson 12-8
___ algebra1.com/extra_examples
___ *Guide to Daily Intervention*, pp. 30–31
___ Daily Intervention, *TWE*, p. 686
___ Study Guide and Intervention, *CRM*, pp. 747–748
___ Reading to Learn Mathematics, *CRM*, p. 751
___ *TeacherWorks CD-ROM*

3 Practice/Apply
___ Skills Practice, *CRM*, p. 749
___ Practice, *CRM*, p. 750
___ Extra Practice, *SE*, p. 848
___ Differentiated Instruction, *TWE*, p. 685
___ *School-to-Career Masters*, p. 24
___ *Parent and Student Study Guide Workbook*, p. 98
___ *Answer Key Transparencies*, Lesson 12-8

Assignment Guide, pp. 686–689, *SE*			
	Objective 1	Objective 2	Other
Basic	11–21 odd	23–33 odd	37–39, 42–68
Average	11–21 odd	23–33 odd	35, 37–39, 42–68
Advanced	12–22 even	24–34 even	36, 40–62 (optional: 63–68)

Alternate Assignment _____

4 Assess
___ Open-Ended Assessment, *TWE*, p. 689
___ Enrichment, *CRM*, p. 752
___ algebra1.com/self_check_quiz

___ *Closing the Gap for Absent Students*, pp. 24–25

KEY *SE* = Student Edition   *TWE* = Teacher Wraparound Edition   *CRM* = Chapter Resource Masters

Lesson Planning Guide (pp. 690–695)

12-9

Teacher's Name _____ Dates _____
Grade _____ Class _____ M Tu W Th F

NCTM Standards
1, 2, 6, 8, 9, 10

Recommended Pacing	
Regular Average	Days 11 & 12 of 14
Regular Advanced	Days 14 & 15 of 17
Block Average	Day 6 of 7
Block Advanced	Day 8 of 9

Objectives
___ Solve rational equations.
___ Eliminate extraneous solutions.
___ State/local objectives: _____

1 Focus
Materials/Resources Needed _____
___ *5-Minute Check Transparencies*, Lesson 12-9
___ Mathematical Background, *TWE*, p. 640D
___ *TeacherWorks CD-ROM*

2 Teach
___ In-Class Examples, *TWE*, pp. 691–693
___ *Teaching Algebra with Manipulatives*, p. 198
___ *Interactive Chalkboard CD-ROM*, Lesson 12-9
___ algebra1.com/extra_examples
___ *Guide to Daily Intervention*, pp. 30–31
___ Daily Intervention, *TWE*, p. 693
___ Study Guide and Intervention, *CRM*, pp. 753–754
___ Reading to Learn Mathematics, *CRM*, p. 757
___ *TeacherWorks CD-ROM*
___ *Multimedia Applications Masters*

3 Practice/Apply
___ Skills Practice, *CRM*, p. 755
___ Practice, *CRM*, p. 756
___ Extra Practice, *SE*, p. 849
___ Differentiated Instruction, *TWE*, p. 692
___ *Graphing Calculator and Spreadsheet Masters*, p. 45
___ *Parent and Student Study Guide Workbook*, p. 99
___ *Answer Key Transparencies*, Lesson 12-9
___ algebra1.com/webquest

Assignment Guide, pp. 693–695, *SE*			
	Objective 1	Objective 2	Other
Basic	11–25 odd	21–25 odd	29–31, 35–48
Average	11–27 odd	21–27 odd	29–31, 35–48
Advanced	12–26 even	20–28 even	32–48
Alternate Assignment			

4 Assess
___ Open-Ended Assessment, *TWE*, p. 695
___ Enrichment, *CRM*, p. 758
___ Assessment, Quiz, *CRM*, p. 774
___ algebra1.com/self_check_quiz

___ Closing the Gap for Absent Students, pp. 24–25

KEY *SE* = Student Edition   *TWE* = Teacher Wraparound Edition   *CRM* = Chapter Resource Masters

© Glencoe/McGraw-Hill   Glencoe Algebra 1

Review and Testing (pp. 696–703)

12

Teacher's Name _____ Dates _____

Grade _____ Class _____ M Tu W Th F

Recommended Pacing	
Regular Average	Days 13 & 14 of 14
Regular Advanced	Days 16 & 17 of 17
Block Average	Day 7 of 7
Block Advanced	Day 9 of 9

Assess

___ *Parent and Student Study Guide Workbook*, p. 100
___ Vocabulary and Concept Check, *SE*, p. 696
___ Vocabulary Test, *CRM*, p. 772
___ Lesson-by-Lesson Review, *SE*, pp. 696–700
___ Practice Test, *SE*, p. 701
___ Chapter 12 Tests, *CRM*, pp. 759–770
___ Open-Ended Assessment, *CRM*, p. 771
___ Standardized Test Practice, *SE*, pp. 702–703
___ Standardized Test Practice, *CRM*, pp. 777–778
___ Cumulative Review, *CRM*, p. 776
___ *Vocabulary PuzzleMaker CD-ROM*
___ algebra1.com/vocabulary_review
___ algebra1.com/chapter_test
___ algebra1.com/standardized_test
___ *MindJogger Videoquizzes VHS*
___ Unit 4 Test, *CRM*, pp. 779–780

Other Assessment Materials

- *TestCheck and Worksheet Builder CD-ROM*

KEY   *SE* = Student Edition   *TWE* = Teacher Wraparound Edition   *CRM* = Chapter Resource Masters

Lesson Planning Guide (pp. 708–714)

13-1

Teacher's Name _____ Dates _____
Grade _____ Class _____ M Tu W Th F

NCTM Standards
1, 5, 6, 8, 9

Recommended Pacing	
Regular Average	Optional
Regular Advanced	Days 1 & 2 of 13
Block Average	Optional
Block Advanced	Days 1 & 2 of 7

Objectives
___ Identify various sampling techniques.
___ Recognize a biased sample.
___ State/local objectives: _____

1 Focus
Materials/Resources Needed _____
___ *5-Minute Check Transparencies*, Lesson 13-1
___ Mathematical Background, *TWE*, p. 706C
___ *TeacherWorks CD-ROM*

2 Teach
___ In-Class Examples, *TWE*, pp. 709–710
___ *Interactive Chalkboard CD-ROM*, Lesson 13-1
___ algebra1.com/extra_examples
___ *Guide to Daily Intervention*, pp. 32–33
___ Study Guide and Intervention, *CRM*, pp. 781–782
___ Reading to Learn Mathematics, *CRM*, p. 785
___ *TeacherWorks CD-ROM*
___ Reading Mathematics, *SE*, p. 714

3 Practice/Apply
___ Skills Practice, *CRM*, p. 783
___ Practice, *CRM*, p. 784
___ Extra Practice, *SE*, p. 849
___ Differentiated Instruction, *TWE*, p. 710
___ Parent and Student Study Guide Workbook, p. 101
___ Answer Key Transparencies, Lesson 13-1

Assignment Guide, pp. 711–713, *SE*			
	Objective 1	Objective 2	Other
Basic	9–21 odd, 29–51	9–21 odd, 29–51	
Average	9–21 odd, 22, 23, 29–51	9–21 odd, 22, 23, 29–51	
Advanced	8–20 even, 22–45 (optional: 46–51)	8–20 even, 22–45 (optional: 46–51)	
Reading Mathematics	1–3		
Alternate Assignment			

4 Assess
___ Open-Ended Assessment, *TWE*, p. 713
___ Enrichment, *CRM*, p. 786
___ algebra1.com/self_check_quiz

___ *Closing the Gap for Absent Students*, pp. 26–27

KEY *SE* = Student Edition   *TWE* = Teacher Wraparound Edition   *CRM* = Chapter Resource Masters

© Glencoe/McGraw-Hill   Glencoe Algebra 1

Lesson Planning Guide (pp. 715–721)

13-2

Teacher's Name _____ Dates _____
Grade _____ Class _____ M Tu W Th F

NCTM Standards
1, 5, 6, 8, 9, 10

Recommended Pacing	
Regular Average	Optional
Regular Advanced	Days 3 & 4 of 13
Block Average	Optional
Block Advanced	Days 2 & 3 of 7

Objectives
___ Organize data in matrices.
___ Solve problems by adding or subtracting matrices or by multiplying by a scalar.
___ State/local objectives: _____

1 Focus
Materials/Resources Needed _____
_____
___ Building on Prior Knowledge, *TWE*, p. 715
___ *5-Minute Check Transparencies*, Lesson 13-2
___ Mathematical Background, *TWE*, p. 706C
___ *TeacherWorks CD-ROM*

2 Teach
___ In-Class Examples, *TWE*, pp. 716–717
___ *Interactive Chalkboard CD-ROM*, Lesson 13-2
___ algebra1.com/extra_examples
___ *Guide to Daily Intervention*, pp. 32–33
___ Daily Intervention, *TWE*, p. 717
___ Study Guide and Intervention, *CRM*, pp. 787–788
___ Reading to Learn Mathematics, *CRM*, p. 791
___ *TeacherWorks CD-ROM*

3 Practice/Apply
___ Skills Practice, *CRM*, p. 789
___ Practice, *CRM*, p. 790
___ Extra Practice, *SE*, p. 849
___ Differentiated Instruction, *TWE*, p. 720
___ *Graphing Calculator and Spreadsheet Masters*, pp. 47, 74
___ *Parent and Student Study Guide Workbook*, p. 102
___ Answer Key Transparencies, Lesson 13-2

Assignment Guide, pp. 718–721, *SE*			
	Objective 1	Objective 2	Other
Basic	17–23 odd, 39	27–33 odd, 40, 41	49–52, 58–70
Average	17–25 odd, 39, 42	27–37 odd, 40, 41, 43, 48	49–52, 58–70 (optional: 53–57)
Advanced	18–26 even, 42, 45, 46	28–38 even, 43, 44, 47, 48	49–68 (optional: 69, 70)
All	Practice Quiz 1 (1–5)		
Alternate Assignment	_____		

4 Assess
___ Practice Quiz 1, *SE*, p. 721
___ Open-Ended Assessment, *TWE*, p. 721
___ Enrichment, *CRM*, p. 792
___ Assessment, Quiz, *CRM*, p. 825
___ algebra1.com/self_check_quiz

___ Closing the Gap for Absent Students, pp. 26–27

KEY *SE* = Student Edition *TWE* = Teacher Wraparound Edition *CRM* = Chapter Resource Masters

© Glencoe/McGraw-Hill Glencoe Algebra 1

Lesson Planning Guide (pp. 722–728)

13-3

Teacher's Name _____ Dates _____

Grade _____ Class _____ M Tu W Th F

NCTM Standards
1, 5, 6, 8, 9, 10

Recommended Pacing	
Regular Average	Optional
Regular Advanced	Day 5 of 13
Block Average	Optional
Block Advanced	Day 3 of 7

Objectives
____ Interpret data displayed in histograms.
____ Display data in histograms.
____ State/local objectives: _____

---

1 Focus
Materials/Resources Needed _____

____ *5-Minute Check Transparencies*, Lesson 13-3
____ *Mathematical Background, TWE*, p. 706D
____ *Prerequisite Skills Masters*, pp. 97–98
____ *TeacherWorks CD-ROM*

2 Teach
____ In-Class Examples, *TWE*, pp. 723–724
____ *Teaching Algebra with Manipulatives*, pp. 200–202
____ *Interactive Chalkboard CD-ROM*, Lesson 13-3
____ algebra1.com/extra_examples
____ algebra1.com/data_update
____ *Guide to Daily Intervention*, pp. 32–33
____ Daily Intervention, *TWE*, p. 725
____ *Study Guide and Intervention, CRM*, pp. 793–794
____ *Reading to Learn Mathematics, CRM*, p. 797
____ *TeacherWorks CD-ROM*

3 Practice/Apply
____ *Skills Practice, CRM*, p. 795
____ *Practice, CRM*, p. 796
____ *Extra Practice, SE*, p. 850
____ *Differentiated Instruction, TWE*, p. 724
____ *School-to-Career Masters*, p. 25
____ *Graphing Calculator and Spreadsheet Masters*, pp. 48, 75
____ *Parent and Student Study Guide Workbook*, p. 103
____ *Answer Key Transparencies*, Lesson 13-3

Assignment Guide, pp. 725–728, SE			
	Objective 1	Objective 2	Other
Basic	11	15–17, 22	23–25, 30–43
Average	11, 13	15, 17, 21	22–25, 30–43 (optional: 26–29)
Advanced	10, 12	14, 18–22 even	19, 21, 23–39 (optional: 40–43)

Alternate Assignment _____

4 Assess
____ Open-Ended Assessment, *TWE*, p. 728
____ Enrichment, *CRM*, p. 798
____ Assessment, Mid-Chapter Test, *CRM*, p. 827
____ Assessment, Quiz, *CRM*, p. 825
____ algebra1.com/self_check_quiz

____ *Closing the Gap for Absent Students*, pp. 26–27

KEY *SE* = Student Edition   *TWE* = Teacher Wraparound Edition   *CRM* = Chapter Resource Masters

Graphing Calculator Investigation (pp. 729-730)
A Follow-Up of Lesson 13-3

13

Teacher's Name _____ Dates _____

Grade _____ Class _____ M Tu W Th F

NCTM Standards
5, 6, 8, 9

Recommended Pacing	
Regular Average	Optional
Regular Advanced	Day 6 of 13
Block Average	Optional
Block Advanced	Day 4 of 7

Objectives
___ Use a graphing calculator to find an appropriate regression equation.
___ State/local objectives: _____

Getting Started
Materials/Resources Needed _____
_____
_____

Teach
___ *Graphing Calculator and Spreadsheet Masters*, p. 76
___ algebra1.com/other_calculator_keystrokes

Assignment Guide, p. 730, *SE*	
All	1–9
Alternate Assignment	

Assess

KEY   *SE* = Student Edition   *TWE* = Teacher Wraparound Edition   *CRM* = Chapter Resource Masters

© Glencoe/McGraw-Hill   Glencoe Algebra 1

Lesson Planning Guide (pp. 731–736)

13-4

Teacher's Name _____ Dates _____

Grade _____ Class _____ M Tu W Th F

NCTM Standards
1, 5, 6, 8, 9, 10

Recommended Pacing	
Regular Average	Optional
Regular Advanced	Days 7 & 8 of 13
Block Average	Optional
Block Advanced	Days 4 & 5 of 7

Objectives
___ Find the range of a set of data.
___ Find the quartiles and interquartile range of a set of data.
___ State/local objectives: _____

---

1 Focus
Materials/Resources Needed _____
_____
___ Building on Prior Knowledge, *TWE*, p. 732
___ 5-Minute Check Transparencies, Lesson 13-4
___ Mathematical Background, *TWE*, p. 706D
___ Prerequisite Skills Masters, pp. 1–2, 19–20
___ TeacherWorks CD-ROM

2 Teach
___ In-Class Examples, *TWE*, pp. 732–733
___ Interactive Chalkboard CD-ROM, Lesson 13-4
___ algebra1.com/extra_examples
___ Guide to Daily Intervention, pp. 32–33
___ Daily Intervention, *TWE*, pp. 732, 733
___ Study Guide and Intervention, *CRM*, pp. 799–800
___ Reading to Learn Mathematics, *CRM*, p. 803
___ TeacherWorks CD-ROM

3 Practice/Apply
___ Skills Practice, *CRM*, p. 801
___ Practice, *CRM*, p. 802
___ Extra Practice, *SE*, p. 850
___ Differentiated Instruction, *TWE*, p. 733
___ Parent and Student Study Guide Workbook, p. 104
___ Real-World Transparency and Master
___ Answer Key Transparencies, Lesson 13-4

Assignment Guide, pp. 734–736, *SE*			
	Objective 1	Objective 2	Other
Basic	19	20–23	11–17 odd, 34–47
Average	24	25–28	11–17 odd, 34–47
Advanced	29	30–32	12–18 even, 33–44 (optional: 45–47)
All	Practice Quiz 2 (1–5)		
Alternate Assignment			

4 Assess
___ Practice Quiz 2, *SE*, p. 736
___ Open-Ended Assessment, *TWE*, p. 736
___ Enrichment, *CRM*, p. 804
___ algebra1.com/self_check_quiz

___ Closing the Gap for Absent Students, pp. 26–27

KEY *SE* = Student Edition   *TWE* = Teacher Wraparound Edition   *CRM* = Chapter Resource Masters

© Glencoe/McGraw-Hill   Glencoe Algebra 1

Lesson Planning Guide (pp. 737–742)

13-5

Teacher's Name _____ Dates _____
Grade _____ Class _____ M Tu W Th F

NCTM Standards
1, 5, 6, 8, 9, 10

Recommended Pacing	
Regular Average	Optional
Regular Advanced	Days 9 & 10 of 13
Block Average	Optional
Block Advanced	Day 5 of 7

Objectives
___ Organize and use data in box-and-whisker plots.
___ Organize and use data in parallel box-and-whisker plots.
___ State/local objectives: _____

1 Focus
Materials/Resources Needed _____

___ Building on Prior Knowledge, *TWE*, p. 737
___ 5-Minute Check Transparencies, Lesson 13-5
___ Mathematical Background, *TWE*, p. 706D
___ TeacherWorks CD-ROM

2 Teach
___ In-Class Examples, *TWE*, pp. 738–739
___ Interactive Chalkboard CD-ROM, Lesson 13-5
___ algebra1.com/extra_examples
___ Guide to Daily Intervention, pp. 32–33
___ Daily Intervention, *TWE*, p. 738
___ Study Guide and Intervention, *CRM*, pp. 805–806
___ Reading to Learn Mathematics, *CRM*, p. 809
___ TeacherWorks CD-ROM

3 Practice/Apply
___ Skills Practice, *CRM*, p. 807
___ Practice, *CRM*, p. 808
___ Extra Practice, *SE*, p. 850
___ Differentiated Instruction, *TWE*, p. 738
___ School-to-Career Masters, p. 26
___ Science and Mathematics Lab Manual, pp. 105–110
___ Parent and Student Study Guide Workbook, p. 105
___ Answer Key Transparencies, Lesson 13-5
___ AlgePASS CD-ROM, Lesson 34
___ algebra1.com/webquest

Assignment Guide, pp. 740–742, *SE*			
	Objective 1	Objective 2	Other
Basic	11, 15–19 odd, 28–31, 39	21–25 odd	40–57
Average	11–19 odd, 30–35, 39	21–27 odd	40–57
Advanced	10–18 even, 32–39	20–26 even	40–57
Alternate Assignment			

4 Assess
___ Open-Ended Assessment, *TWE*, p. 742
___ Enrichment, *CRM*, p. 810
___ Assessment, Quiz, *CRM*, p. 826
___ algebra1.com/self_check_quiz

___ Closing the Gap for Absent Students, pp. 26–27

KEY   *SE* = Student Edition   *TWE* = Teacher Wraparound Edition   *CRM* = Chapter Resource Masters

© Glencoe/McGraw-Hill   Glencoe Algebra 1

Algebra Activity (pp. 743–744)
A Follow-Up of Lesson 13-5

13

Teacher's Name _____ Dates _____
Grade _____ Class _____ M Tu W Th F

NCTM Standards
1, 5, 6, 9, 10

Recommended Pacing	
Regular Average	Optional
Regular Advanced	Day 11 of 13
Block Average	Optional
Block Advanced	Day 6 of 7

Objectives
___ Use tables to determine percentiles.
___ State/local objectives: _____

Getting Started
Materials/Resources Needed paper, pencil, ruler
_____
_____

Teach
___ *Teaching Algebra with Manipulatives*, p. 203
___ *Glencoe Mathematics Classroom Manipulative Kit*
___ Teaching Strategy, *TWE*, p. 743

Assignment Guide, pp. 743–744, *SE*	
All	1–10
Alternate Assignment	

Assess
___ Study Notebook, *TWE*, p. 744

KEY *SE* = Student Edition   *TWE* = Teacher Wraparound Edition   *CRM* = Chapter Resource Masters

© Glencoe/McGraw-Hill   Glencoe Algebra 1

Review and Testing (pp. 745–751)

13

Teacher's Name _____ Dates _____
Grade _____ Class _____ M Tu W Th F

Recommended Pacing	
Regular Average	Optional
Regular Advanced	Days 12 & 13 of 13
Block Average	Optional
Block Advanced	Day 7 of 7

Assess

___ *Parent and Student Study Guide Workbook*, p. 106
___ Vocabulary and Concept Check, *SE*, p. 745
___ Vocabulary Test, *CRM*, p. 824
___ Lesson-by-Lesson Review, *SE*, pp. 745–748
___ Practice Test, *SE*, p. 749
___ Chapter 13 Tests, *CRM*, pp. 811–822
___ Open-Ended Assessment, *CRM*, p. 823
___ Standardized Test Practice, *SE*, pp. 750–751
___ Standardized Test Practice, *CRM*, pp. 829–830
___ Cumulative Review, *CRM*, p. 828
___ *Vocabulary PuzzleMaker CD-ROM*
___ algebra1.com/vocabulary_review
___ algebra1.com/chapter_test
___ algebra1.com/standardized_test
___ *MindJogger Videoquizzes VHS*

Other Assessment Materials

- *TestCheck and Worksheet Builder CD–ROM*

KEY *SE* = Student Edition   *TWE* = Teacher Wraparound Edition   *CRM* = Chapter Resource Masters

© Glencoe/McGraw-Hill   Glencoe Algebra 1

Lesson Planning Guide (pp. 754–758)

14-1

Teacher's Name _____ Dates _____
Grade _____ Class _____ M Tu W Th F

NCTM Standards
1, 5, 6, 8, 9, 10

Recommended Pacing	
Regular Average	Optional
Regular Advanced	Day 1 of 11
Block Average	Optional
Block Advanced	Day 1 of 5.5

Objectives
___ Count outcomes using a tree diagram.
___ Count outcomes using the Fundamental Counting Principle.
___ State/local objectives: _____

---

1 Focus
Materials/Resources Needed _____

___ *5-Minute Check Transparencies*, Lesson 14-1
___ Mathematical Background, *TWE*, p. 752C
___ *TeacherWorks CD-ROM*

2 Teach
___ In-Class Examples, *TWE*, pp. 755–756
___ *Interactive Chalkboard CD-ROM*, Lesson 14-1
___ algebra1.com/extra_examples
___ *Guide to Daily Intervention*, pp. 34–35
___ Study Guide and Intervention, *CRM*, pp. 831–832
___ Reading to Learn Mathematics, *CRM*, p. 835
___ *TeacherWorks CD-ROM*

3 Practice/Apply
___ Skills Practice, *CRM*, p. 833
___ Practice, *CRM*, p. 834
___ Extra Practice, *SE*, p. 851
___ Differentiated Instruction, *TWE*, p. 755
___ *Parent and Student Study Guide Workbook*, p. 107
___ Answer Key Transparencies, Lesson 14-1

Assignment Guide, pp. 756–758, *SE*			
	Objective 1	Objective 2	Other
Basic	9	15, 17	11, 13, 23–51
Average	9, 19	15, 17	11, 13, 18, 23–51
Advanced	10	16, 20–22	12, 14, 23–45 (optional: 46–51)
Alternate Assignment			

4 Assess
___ Open-Ended Assessment, *TWE*, p. 758
___ Enrichment, *CRM*, p. 836
___ algebra1.com/self_check_quiz

___ Closing the Gap for Absent Students, pp. 28–29

KEY *SE* = Student Edition   *TWE* = Teacher Wraparound Edition   *CRM* = Chapter Resource Masters

© Glencoe/McGraw-Hill   Glencoe Algebra 1

Algebra Activity (p. 759)
A Follow-Up of Lesson 14-1

14

Teacher's Name _____ Dates _____

Grade _____ Class _____ M Tu W Th F

NCTM Standards
3, 5, 6, 7, 8, 9, 10

Recommended Pacing	
Regular Average	Optional
Regular Advanced	Day 1 of 11
Block Average	Optional
Block Advanced	Day 1 of 5.5

Objectives
____ Use finite graphs to determine whether a route is traceable.
____ State/local objectives: _____

Getting Started
Materials/Resources Needed paper, pencil
_____
_____

Teach
____ *Teaching Algebra with Manipulatives*, p. 206
____ *Glencoe Mathematics Classroom Manipulative Kit*
____ *Teaching Strategy, TWE*, p. 759

Assignment Guide, p. 759, *SE*	
All	1–8
Alternate Assignment	

Assess
____ Study Notebook, *TWE*, p. 759

KEY	*SE* = Student Edition	*TWE* = Teacher Wraparound Edition	*CRM* = Chapter Resource Masters

© Glencoe/McGraw-Hill   Glencoe Algebra 1

Lesson Planning Guide (pp. 760–768)

14-2

Teacher's Name _____ Dates _____

Grade _____ Class _____ M Tu W Th F

NCTM Standards
1, 5, 6, 8, 9, 10

Recommended Pacing	
Regular Average	Optional
Regular Advanced	Days 2 & 3 of 11
Block Average	Optional
Block Advanced	Days 1 & 2 of 5.5

Objectives
___ Determine probabilities using permutations.
___ Determine probabilities using combinations.
___ State/local objectives: _____

1 Focus
Materials/Resources Needed _____

___ *5-Minute Check Transparencies*, Lesson 14-2
___ Mathematical Background, *TWE*, p. 752C
___ *TeacherWorks CD-ROM*

2 Teach
___ In-Class Examples, *TWE*, pp. 761–763
___ *Interactive Chalkboard CD-ROM*, Lesson 14-2
___ algebra1.com/extra_examples
___ algebra1.com/data_update
___ *Guide to Daily Intervention*, pp. 34–35
___ Daily Intervention, *TWE*, p. 764
___ Study Guide and Intervention, *CRM*, pp. 837–838
___ Reading to Learn Mathematics, *CRM*, p. 841
___ *TeacherWorks CD-ROM*
___ Reading Mathematics, *SE*, p. 768

3 Practice/Apply
___ Skills Practice, *CRM*, p. 839
___ Practice, *CRM*, p. 840
___ Extra Practice, *SE*, p. 851
___ Differentiated Instruction, *TWE*, p. 761
___ School-to-Career Masters, p. 27
___ Parent and Student Study Guide Workbook, p. 108
___ Answer Key Transparencies, Lesson 14-2
___ algebra1.com/webquest

Assignment Guide, pp. 764–767, SE			
	Objective 1	Objective 2	Other
Basic	15–21 odd, 23, 27–31 odd	15–21 odd, 25, 27	33–39, 50, 51, 55–77
Average	15–21 odd, 23, 29–31 odd	15–21 odd, 25, 27	33, 36–44, 50, 51, 55–77
Advanced	14–20 even 22, 28, 30	14–20 even, 24, 26, 32	43–71 (optional: 72–77)
All	Practice Quiz 1 (1–5)		
Reading Mathematics	1–4		
Alternate Assignment			

4 Assess
___ Practice Quiz 1, *SE*, p. 767
___ Open-Ended Assessment, *TWE*, p. 767
___ Enrichment, *CRM*, p. 842

___ Assessment, Quiz, *CRM*, p. 875
___ algebra1.com/self_check_quiz

___ *Closing the Gap for Absent Students*, pp. 28–29

KEY   *SE* = Student Edition   *TWE* = Teacher Wraparound Edition   *CRM* = Chapter Resource Masters

© Glencoe/McGraw-Hill   Glencoe Algebra 1

Lesson Planning Guide (pp. 769–776)

14-3

Teacher's Name _____ Dates _____

Grade _____ Class _____ M Tu W Th F

NCTM Standards
1, 5, 6, 8, 9, 10

Recommended Pacing	
Regular Average	Optional
Regular Advanced	Days 4 & 5 of 11
Block Average	Optional
Block Advanced	Days 2 & 3 of 5.5

Objectives

____ Find the probability of two independent events or dependent events.
____ Find the probability of two mutually exclusive or inclusive events.
____ State/local objectives: _____

1 Focus

Materials/Resources Needed _____
_____

____ *5-Minute Check Transparencies*, Lesson 14-3
____ *Mathematical Background, TWE*, p. 752D
____ *Prerequisite Skills Masters*, pp. 47–48, 55–56, 67–70, 99–100
____ *TeacherWorks CD-ROM*

2 Teach

____ In-Class Examples, *TWE*, pp. 770–772
____ *Teaching Algebra with Manipulatives*, p. 207
____ *Interactive Chalkboard CD-ROM*, Lesson 14-3
____ algebra1.com/extra_examples
____ *Guide to Daily Intervention*, pp. 34–35
____ Daily Intervention, *TWE*, p. 773
____ Study Guide and Intervention, *CRM*, pp. 843–844
____ Reading to Learn Mathematics, *CRM*, p. 847
____ *TeacherWorks CD-ROM*

3 Practice/Apply

____ Skills Practice, *CRM*, p. 845
____ Practice, *CRM*, p. 846
____ Extra Practice, *SE*, p. 851
____ Differentiated Instruction, *TWE*, p. 770
____ *Science and Mathematics Lab Manual*, pp. 111–116
____ *Parent and Student Study Guide Workbook*, p. 109
____ *Real-World Transparency and Master*
____ *Answer Key Transparencies*, Lesson 14-3
____ *AlgePASS CD-ROM*, Lesson 35

Assignment Guide, pp. 772–776, SE			
	Objective 1	Objective 2	Other
Basic	17–23 odd, 28–34	25, 27	35, 48–75
Average	17–23 odd, 36–40	25, 27	35, 48–75
Advanced	16–22 even, 38–40	24, 26, 41–47	48–66 (optional: 67–75)
Alternate Assignment			

4 Assess

____ Open-Ended Assessment, *TWE*, p. 776
____ Enrichment, *CRM*, p. 848
____ Assessment, Mid-Chapter Test, *CRM*, p. 877
____ Assessment, Quiz, *CRM*, p. 875
____ algebra1.com/self_check_quiz

____ *Closing the Gap for Absent Students*, pp. 28–29

KEY *SE* = Student Edition   *TWE* = Teacher Wraparound Edition   *CRM* = Chapter Resource Masters

Lesson Planning Guide (pp. 777–781)

14-4

Teacher's Name _____ Dates _____
Grade _____ Class _____ M Tu W Th F

NCTM Standards
1, 5, 6, 8, 9, 10

Recommended Pacing	
Regular Average	Optional
Regular Advanced	Days 6 & 7 of 11
Block Average	Optional
Block Advanced	Days 3 & 4 of 5.5

Objectives
____ Use random variables to compute probability.
____ Use probability distributions to solve real-world problems.
____ State/local objectives: _____

1 Focus
Materials/Resources Needed _____
____ *5-Minute Check Transparencies*, Lesson 14-4
____ Mathematical Background, *TWE*, p. 752D
____ *TeacherWorks CD-ROM*

2 Teach
____ In-Class Examples, *TWE*, p. 778
____ *Interactive Chalkboard CD-ROM*, Lesson 14-4
____ algebra1.com/extra_examples
____ *Guide to Daily Intervention*, pp. 34–35
____ Study Guide and Intervention, *CRM*, pp. 849–850
____ Reading to Learn Mathematics, *CRM*, p. 853
____ *TeacherWorks CD-ROM*

3 Practice/Apply
____ Skills Practice, *CRM*, p. 851
____ Practice, *CRM*, p. 852
____ Extra Practice, *SE*, p. 852
____ Differentiated Instruction, *TWE*, p. 778
____ *Graphing Calculator and Spreadsheet Masters*, p. 50
____ *Parent and Student Study Guide Workbook*, p. 110
____ *Answer Key Transparencies*, Lesson 14-4

Assignment Guide, pp. 779–781, *SE*			
	Objective 1	Objective 2	Other
Basic	10, 11, 14	12, 13, 15–17	23–48
Average	10, 11, 14	12, 13, 15–17	23–48
Advanced	18	18–22	23–42 (optional: 43–48)
All	Practice Quiz 2 (1–5)		
Alternate Assignment			

4 Assess
____ Practice Quiz 2, *SE*, p. 781
____ Open-Ended Assessment, *TWE*, p. 781
____ Enrichment, *CRM*, p. 854
____ Assessment, Quiz, *CRM*, p. 876
____ algebra1.com/self_check_quiz

____ *Closing the Gap for Absent Students*, pp. 28–29

KEY *SE* = Student Edition *TWE* = Teacher Wraparound Edition *CRM* = Chapter Resource Masters

© Glencoe/McGraw-Hill   Glencoe Algebra 1

Lesson Planning Guide (pp. 782–788)

14-5

Teacher's Name _____ Dates _____

Grade _____ Class _____ M Tu W Th F

NCTM Standards
1, 5, 6, 8, 9, 10

Recommended Pacing	
Regular Average	Optional
Regular Advanced	Days 8 & 9 of 11
Block Average	Optional
Block Advanced	Days 4 & 5 of 5.5

Objectives

___ Use theoretical and experimental probability to represent and solve problems involving uncertainty.
___ Perform probability simulations to model real-world situations involving uncertainty.
___ State/local objectives: _____

1 Focus

Materials/Resources Needed _____
_____
___ Building on Prior Knowledge, *TWE*, p. 782
___ 5-Minute Check Transparencies, Lesson 14-5
___ Mathematical Background, *TWE*, p. 752D
___ TeacherWorks CD-ROM

2 Teach

___ In-Class Examples, *TWE*, pp. 783–784
___ *Teaching Algebra with Manipulatives*, pp. 208–209
___ *Interactive Chalkboard CD-ROM*, Lesson 14-5
___ algebra1.com/extra_examples
___ algebra1.com/careers
___ *Guide to Daily Intervention*, pp. 34–35
___ Study Guide and Intervention, *CRM*, pp. 855–856
___ Reading to Learn Mathematics, *CRM*, p. 859
___ TeacherWorks CD-ROM

3 Practice/Apply

___ Skills Practice, *CRM*, p. 857
___ Practice, *CRM*, p. 858
___ Extra Practice, *SE*, p. 852
___ Differentiated Instruction, *TWE*, p. 784
___ *School-to-Career Masters*, p. 28
___ *Graphing Calculator and Spreadsheet Masters*, pp. 49, 77
___ *Parent and Student Study Guide Workbook*, p. 111
___ *Answer Key Transparencies*, Lesson 14-5
___ algebra1.com/webquest

Assignment Guide, pp. 785–788, *SE*			
	Objective 1	Objective 2	Other
Basic	22–24	13–21	32–35, 39–63
Average	22–24	13–21	25–27, 32–35, 39–63 (optional: 36–38)
Advanced	22–24	19–21, 25–31	32–63
Alternate Assignment			

4 Assess

___ Open-Ended Assessment, *TWE*, p. 788
___ Enrichment, *CRM*, p. 860
___ Assessment, Quiz, *CRM*, p. 876
___ algebra1.com/self_check_quiz

___ *Closing the Gap for Absent Students*, pp. 28–29

KEY *SE* = Student Edition   *TWE* = Teacher Wraparound Edition   *CRM* = Chapter Resource Masters

© Glencoe/McGraw-Hill   Glencoe Algebra 1

Review and Testing (pp. 789–795)

14

Teacher's Name _____ Dates _____

Grade _____ Class _____ M Tu W Th F

Recommended Pacing	
Regular Average	Days 12 & 13 of 13
Regular Advanced	Days 13 & 14 of 14
Block Average	Day 7 of 7
Block Advanced	Day 8 of 8

Assess

____ *Parent and Student Study Guide Workbook*, p. 112
____ Vocabulary and Concept Check, *SE*, p. 789
____ Vocabulary Test, *CRM*, p. 874
____ Lesson-by-Lesson Review, *SE*, pp. 789–792
____ Practice Test, *SE*, p. 793
____ Chapter 14 Tests, *CRM*, pp. 861–872
____ Open-Ended Assessment, *CRM*, p. 873
____ Standardized Test Practice, *SE*, pp. 794–795
____ Standardized Test Practice, *CRM*, pp. 879–880
____ Cumulative Review, *CRM*, p. 878
____ *Vocabulary PuzzleMaker CD-ROM*
____ algebra1.com/vocabulary_review
____ algebra1.com/chapter_test
____ algebra1.com/standardized_test
____ *MindJogger Videoquizzes VHS*
____ Unit 5 Test, *CRM*, pp. 881–882
____ Second Semester Test, *CRM*, pp. 883–886
____ Final Test, *CRM*, pp. 887–892

Other Assessment Materials

- *TestCheck and Worksheet Builder CD-ROM*

KEY *SE* = Student Edition   *TWE* = Teacher Wraparound Edition   *CRM* = Chapter Resource Masters